Doomsday Prepping and Survival

From Civil Disturbances to Biblical Proportions

JAIME MERA

Dedication

I dedicate this book to all those people looking to survive a global crisis or apocalypse. This is not rocket science information, and is for anyone who can understand basic concepts, methods, and intent with limited or unlimited resources.

Published books:

Jesus and the Paint on the Wall: What Do People Live For? (2012)

Doomsday Prepping and Survival: From Civil Disturbances to
Biblical Proportions (2014)

**Science Fiction
A Superhero Epic Series**

Creator (2004, 2014)

He Is Known as Ego (2006, 2014)

Guild Without a Name (2014)

The Galaxy Is Ours (2014)

Masterminds (2014)

Preface

* | * | *

T his book is not a list of recipes or inventory for you to pack a bag and survive in the wilderness for a week, a year, or a lifetime. This book covers the simple to worst case scenarios in surviving an apocalyptic event like the complete economic failure of a country to the devastation of a meteor large enough to kill millions to billions of people. It is based on experience, research, logic, and a leap of faith to survive something which is yet to personally and fully reveal its ugly head at you. I am not talking about a hurricane, tornado, out of control fire, flash flood, earthquake, sink holes, or avalanche types of events. I am talking about the collapse of the economic and legal system of a country, a chaotic internal war with nuclear consequences, nationwide dirty bomb effects, super volcano, massive meteor strike, and the events depicted in the bible prior to the seventh trumpet.

This book is for those people who might be interested in surviving an apocalyptic event or who are sure and determined in surviving such events. This is not for the person who does not believe an economic crisis will occur, a depletion of resources is putting us on the verge of extinction, the biblical events of the second coming, or a stellar doom scenario are not possible. I highlight these four events because they are all catastrophic to the point where without preparation you will not survive or come close to surviving. If a war erupted it is possible to survive simply by chance of being in the right place at the right time. However, if you look at biblical accounts, wars and rumors of wars is part of a chain of events. So several events can constitute one major event in logical respects, but it is more than that. A person can survive a conventional war as a Soldier and come home to die of a new plague brought on by the war or by the next door neighbor. Or, a family could survive a world war and die from a civil war because their ethnic background has somehow in someone's mind made them less of a human being. Yes, these things have already happened in history, but the doomsday events which I will talk about have not happened, yet.

Different scenarios will be examined, to include chain of events and practical steps in putting yourself in the best possible position to survive them. This book will not guarantee that you will survive, but brings up questions and solutions to what the average person and experts might think of but dismiss as not important, time consuming, highly unlikely, or not as important as other factors. This book looks at the ABCs, and goes beyond to get you focused on the best chances of success in your endeavor to withstand the storm that is coming.

I started the first chapter with a question and it may seem dry or too simplistic to some people; but you must keep in mind your chances of survival stands on a foundation of a mind set, determination, and chance otherwise known as dumb luck. I call the later, not luck, but God's favor for those who believe in God. I want you to think of every aspect of the questions and answers in this book, and write down or create your own conclusions from logic, your research, and what experts supposedly say. I try to explain things as simple as possible and at the same time, only talk about what is relevant for you or other preppers. For example: I would highly recommend you stock up on medications, it might be relevant only to people with diabetes for instance, but in fact is essential for everyone to stock up on all medications for all contingencies. Not that you have to have exotic meds for a possible psychosis that does not exist at the moment, but you do need meds for pain, seizures, stock, infections, and the like. Take what is said in this book as a compass and adapt to your particular situation.

Contents

Chapter One

* I * I *

WHAT CAN CAUSE ME TO DIE?

Sun spots or solar flares are said to be catastrophic to radio or telephonic communications. But what do events like this really do to us. Well the Sun is one of the top ten sources of our destruction. A solar flare that doesn't actually go near the Earth has the potential of burning all of the green grass on the Earth, which is one event in the bible that will happen according to Revelation 8:7, "And the first (trumpet) sounded, and there came hail and fire, mixed with blood, and they were thrown on the Earth; and a third of the Earth was burned up, and a third of the trees were burned up, and all of the green grass was burned up (NASB, 1990)." The bible doesn't specify exactly how, but it could be a solar flare, meteors, missiles, or some other form we have not yet seen or thought about; either way, the destruction of green grass, trees, and the like will cause us to die. Not that we will die from burning up unless we are outside when the periods of heat occur, but more from starvation. This starvation scenario can occur in many ways from a solar flare, a massive loss of crops due to worldwide or country wide chemicals or mutant insects, destruction of specific ecosystems, or an economic breakdown in society. I will key into the preparations to minimize these effects in a later chapter.

Direct physical harm can also cause us to die. It can be biological in nature, but it is more likely to be physical harm in the sense of a great earthquake, a super volcano, physical water harm, or meteor effects. The other physical type of harm which is likely, depending on the circumstance, is harm from other people. In the middle of a food or economic crisis, other people will be either an ally or enemy to your goals of survival. For many people who depend on medicine to keep them alive, the additional struggle will be to maintain their medical care or find an alternative medical supplement. These are people who need a steady supply of medication, like people with diabetes or seizure patients. There are certain drugs that cannot be stocked up because of legal distribution laws, so when there are no substitutes the people will become a victim to the event because of the dependency to the social and economic architecture.

Another physical method of harm is the effects of a super volcano or meteor strike which are impact, puncture of the body by fragments of all kinds, heat, poison, suffocation, and destruction of the food source. But these are only primary effects, the secondary effects are the same with medication supplies stopping, water and food stopping on an extended basis, and any survivors above ground doing what they think is in their best interest, not yours or your group.

Aside from all of these factors in what will kill you, is yourself. You can physically and emotionally cause yourself to die, slowly or very quickly. Your attitude and reaction to events can and will determine if your chances of survival are in your favor. It is more than being positive and determined in surviving, because you can be extremely determined and get shot by a sniper or accidentally die because you were in the wrong place at the wrong time. Your determination and preparation is a tool, not a guarantee. You and the ones working with you are the force behind that tool.

However, the people with you can also make you die or they may raise your odds of living. Adapting to change is your measuring stick of survival when it comes to yourself and people around you. You must be able to adapt to your environment and the people with you must also be able to adapt. You can have a group of six adults and if one cannot adapt fast enough or doesn't want to adapt, your group may suffer to the point of death. An example of this is if you are forced to be on the move by foot and stop in a defensive position, pulling security. One of your members has not been drilled with this contingency and is also not in the state of mind to adapting to most situations. In this case, you have wounded, and the rotation placed on the group for security is 50% which is not terrible, but not ideal. Your member falls asleep instead of alerting the leader for a relief, or getting on a knee or even standing up to stay awake. The result is your position gets overrun and your group is captured or killed. This can happen in a stationary fortified location, and it could be you that falls asleep or ignores the fact that your communication with the center of the group is gone, because of low batteries on your radio. Whatever, the reasons, the way your group adapts and preps for any event will determine your survival rate.

So back to you being a game changer. If you are a negative person or think you are always right, then you will be your own downfall. A great leader is someone who knows their limitations and is flexible. A follower who ends up hurt or dead is either a hero who has followed the leader's orders to the T, or is someone who ignored safety practices, didn't pay attention, or believes he/she is not going to make it and acts in fear. Fear… it is not a word you should take likely. Fear is not just running away or freezing in your tracks. No, fear is also anxiety or worry about the future and present. Worry that someone may come into your home and start spraying it with bullets because they want to take your food, rob you, or occupy your home for shelter. It is not wrong to fear, but when that is all you do most of the time, then yes it is wrong to constantly fear something which might or might not happen. Your negative attitude and constant fear will more than likely keep you from being a prepper, but if it has not, your negative or fearful attitude will keep you from accepting what really works and ignoring the things you shouldn't waste your time and effort on.

If you think that the economy is going to breakdown and you think it will never recover, then you should move to a different country or plan on surviving independently (being self sufficient) for the rest of your life. Otherwise, think positive and perhaps you will be the one of many people who leads the country back into economic stability. So I must ask those who are negative thinkers, if you don't think positive, then why are you so motivated to prep and survive?

There is a movie that came out around 2011 which I don't want to advertise by naming it because of the message and other flaws, but it starts with a nuclear explosion and a group of people who make it into a shelter of an apartment they live in. The shelter was made by a prepper who was also the superintendent. Apparently another country invaded the USA which is why the bomb went off, and the movie goes into this story about the people in the shelter. Well, it shows the social interactions with the people and in the end all of them die because of greed, stupidly, and lust; except for one girl (now young widow) who escapes through the sewer system and comes up to the street level wearing a chemical (supposedly also radioactive) protective suit. She sees the absolute devastation caused by the nuclear explosion and that is how the movie ends.

So what is the moral of the story? The people in the shelter acted like immoral killers when things didn't go their way. They wanted to survive only if they got their way. But survive to what end? If they saw and knew what was waiting for them on the surface, maybe they would not have wanted to continue living knowing the radiation would kill them slowly. By the way, the radiation came into the shelter when they opened the main door and allowed the dust particles to come in, so half of the people in the shelter were dying of radiation poisoning before they started to kill each other trying to control one another. What is the moral of the story again? If you are going to be negative, then just make it easier on yourself and stand in the middle of ground zero so you don't have to go through a lot of pain and suffering in the end. Or just eat what you can until there is no more and die of starvation.

It is your choice; have a positive attitude, have a plan, and execute it; or be negative and worry about life, have no plan, and perhaps you will see the last days on the wrong end of the viewing glass. Like I said, this book is for those who believe there is hope at the end of the tunnel, not those who let fear run their lives. Having said that, fear can be overcome, and your first step in overcoming that fear is to have faith. I would say have faith in God and His love for you which casts out all fear, but for those who don't believe in God, then have faith in yourself, those you trust around you, and the preparation you put into surviving. Drills, drills, and more drills will help take away or reduce your fears on many levels. Practice makes better, and the greater your confidence is, the less fear you will have.

So, there are many things which can cause you to die. Being in the wrong place at the wrong time, being unprepared, being injured and unprepared, having the wrong people around you, having the wrong attitude – like pride, fear or anxiety, lack of the right resources, prepping for the wrong events, and lack of practice. In the end, there are too many things that you will not be able to foresee, but your preparation will be what helps you adapt and make your chances higher than normal.

I will cover the best scenario for you to be in, but it is also the hardest to attain. Everyone is different and the life which a person lives is in many degrees, destined to happen. You might be born a prince or princess, a child of a wealthy family, or a very poor person who lives in the slums. You might be broke right now wanting to file bankruptcy while going through a divorce. You might have a promising career and make tons of money but have disabled family members you have to care for the rest of their lives. Your life situation in the past and right now was not all under your control. <u>What you do</u> now and in the future will be your choice. You can prep slowly or quickly, or you may not have a choice but to prep slowly. Whatever your situation is, there is hope.

The best case survival scenario will require that you have a lot of resources, the right family or group makeup, and the right plan. I suggest you look at other doomsday survival resources, but only to get what you think best fits you and you will see that the best scenario is very logical, and in the end it should be your final goal. If anything, trying to get to the best case scenario will take whatever course of action you plan on taking, or are already taking to higher levels you never considered or dreamt about. I say this, because there are in fact some decent plans out there created by very motivated preppers, but like in most things, there is room for improvement.

Improvement, not in how a prepper can fire a weapon or take down a hostile to the ground. No, improvement in preparation time and efforts so that it's not wasted even though supposed experts say your plan is good or in some cases terrible. I will address this issue in the next chapter talking about shows like Doomsday Preppers and Doomsday Castle, but keep in mind that many sources like television and books, tend to advertise ideas as if they are absolute fact for the ratings and audience to keep watching by claiming that the experts say this and that giving the impression that the experts are real experts or know what they are talking about.

There was this one episode in Doomsday Preppers where the children at a very young age were taught to fire weapons and they were conducting some small unit tactic drills. They focused on hostage drills, and the children under 16 years of age learned well for their age, but there is a big difference with a professional SWAT team or Special Operations Soldier teaching you drills, than your father who was raised with weapons. There is merit to the father being very knowledgeable and proficient in weapons; but when you have a squad size element of adults or at the very least age 15 and above who have been trained by infantry veteran Soldiers or better; and they run the younger kids through the training, then the training will be worth its weight in gold. The adults will also get more out of the training and your plan will be better. Granted, the father got aid from the local police, which helped out a little, but there were too many things left out which the local police were and are simply not experienced or knowledgeable on.

Police officers are knowledgeable to a point, but unless they have been in a SWAT team, commanded similar teams, or had similar training, they would not know how to effectively counter a sniper or an organized assault team in the open, in an indoor complex, or sewer system. If there is an economic disaster and I were a qualified sniper with my own family or band of brothers, than chances are that I would not go around and kill people or families just to take the bounty of their hard work for survival. But, if I am sniper qualified and sadly to say want to go wild and do evil, then it is highly unlikely the prepper would know how to deal with a sniper picking you off if you have a safe haven above ground or go out in the open. I talk about a sniper, because it's one of the easiest concept for someone to understand when it comes to thinking you are safe in your home with bullet proof walls and hopefully windows too; but what happens when a person approaches the house? Close enough to throw a grenade or launch a grenade at your home. Do you go outside and challenge them, try to shoot them through an opened window or door? Shoot them through a firing port? Once you go outside or open the door or window, a sniper takes you out. What then?

There will hopefully be some law and order in the country and you don't have to worry about a tank coming up to your safe area and blowing up your main house or bunker, so they can take your water supply from a nearby well. If a super volcano lays waste to the land, chances are there will be no tanks, or weapons that can cause that much damage to you or anything you occupy. But there are weapons that can cause a lot of damage which a person can carry, like a sniper rifle, grenade launcher, or worse (an illegally made flame thrower). So, what I'm trying to say is don't take information from shows like Doomsday Preppers as fact just because the experts analyzed the preppers' chances to be high or low. Take everything you see and put it into context, use some logic, and then ask yourself if it were you doing that, would you do it differently? Then conduct your own research and start prepping with a plan and goal in mind.

You and people like you are what can keep you alive, apart from all that can kill you. It is very important to note, and I will talk about this in great detail in later chapters, but surviving alone is not what you want or should want to do. It can happen, but if for some reason you happen to be the only survivor on your continent with no means of communicating with any other survivors around the world or capable of traveling thousands of miles to where they happen to be, then all you did was get yourself a front courtside seat to your lonely life, however long that may last.

The last thing I will point out is the fact that this world is a fallen world, which means it is cursed with death in the form of aging, disease, or whatever microscopic killer you want to name. In the end, biological aspects of death is something all preppers need to survive against in all scenarios. I will talk later about livestock, disease, pets, and a few more things which link to your survival against the unseen.

Chapter Two

✳ ❙ ✳ ❙ ✳

ALL THOSE SURVIVAL THINGS WE SEE AND READ ABOUT

The struggle to survive has been embraced by the idea that you have to get into top physical shape, be educated, be skilled, and/or have the right equipment. Unfortunately, the idea is partially correct. You cannot survive a hunger crisis if you are malnourished or extremely over weight. If you are extremely over weight, you might last a long time without food, but depending on your situation, you will not last long if you have to move around a lot or use up a lot more water than normal. Being in good physical shape will help greatly in conserving water and food. It will also allow you to physically meet the demands of staying in shape while in an enclosed space, fighting off enemies, or traveling long distances on foot or by bike. Don't misunderstand, being in top physical shape is not being able to run five or ten miles at a six and a half minute mile pace, or doing fifty correct pushups in less than a minute. The average fit person should be able to walk four miles in an hour and not be exhausted, in fact the person should be able to walk twenty miles in five hours without carrying a heavy load, and not be exhausted to the point of being useless and unable to pitch a tent or think clearly. It takes practice and determination. There are many people who think four miles an hour is too slow or fast, but if survival is your goal, a constant moderate speed is always going to prevail over a quick exhausting speed. If you are too tired to fight or do manual labor at the end of your travels, then you have failed in staying in shape. Also, if you can lift objects over 20 lbs, and can pick up at least half your body weight you are in a position to survive; however, it would help greatly if you could pick up your entire body weight in case a lot of climbing is involved for your survival like being near mountains or many built up areas like a city.

There is a lot of information out there about survival in general and not so much about surviving an epic doomsday scenario, but even then much of the information is scattered. The problem is that the information will always be scattered because there are different levels of expertise, survival techniques, scenarios, and resources. I will try to point you in the right direction and let you decide what areas you want to focus your educational efforts. A quick snap shot of educational resources is actually on cable television, not just in the internet. Discovery, History, Military Channel and a few more like real TV, actually expose you to doomsday documentaries, Doomsday Preppers type of shows, building underground, Myth Busters, and valuable innovations like the Nu-wave oven, vacuum packing systems, or Little Giant ladder. You just need to look at what to do with all that information. You can get all sorts of books on making a fire with strings and a stick, trapping animals, hunting, eating off of the land, or making shelters; but all of this only helps you to survive in the woods, desert, or swamps; not in an ash covered land, or diseased and ash covered city. I use ash as an example because a meteor(s) or super volcano will cause ash to fall on the Earth. If you happen to survive the initial destructive effects, you will be in a land covered by ash and dead flesh. So, to be blunt; studying about surviving in the wilderness will only help if there is an economic crisis, otherwise hunting in ash or radioactive covered lands for dead animals is not a skill I would want to know since there are other things that will be more important. Hunting is something you do want to know for a post doomsday situation, and it should be at the bottom of the list to learn or do. In fact farming should be higher in the list since your chances of planting seeds that you have stored will probably be easier to do than you finding living animals in a land devastated by debris, dead vegetation and corpses.

Knowledge is power and in many ways, knowledge is survival, but not as people think. What you know can make life easier and save you, but what you don't know can also kill you. There are people who have made elaborate life styles in growing food in their backyard or in an underground room. They are very knowledgeable about nature and have a year or more of food supply. They even have waste disposal systems and hygiene resources that can sustain them for years. Many understand the need for security, but there are some that don't want to use weapons, in particular firearms, and have 'no weapon policies'. That is noble and all, but what many people fail to understand is the noble idea of not willing to defend yourself with a firearm is a major favorable factor in another person's determination to take what you have, because they are desperate or only care about themselves at your expense, since they were not prepping as you were, but enjoying the life they had in the good old days. These people will more than likely have a weapon from a dead police officer, a neighbor, their own home, a gun shop; just about anywhere if you look hard enough when a doomsday crisis has occurred. They might not know how exactly to use the weapon, but it is not that hard to learn when you are in a crisis situation. If they do know how to use the firearm, then it's likely they will be accurate and deadly with it.

Now, why have I placed so much importance on a firearm in the hands of someone who might use one against you? Well, if you are very knowledgeable in firearms, you would know what weapons are most deadly, which ones are easiest to shoot, are the most reliable, can penetrate body armor, are easiest to conceal, will load you down with ammunition, and which ones will take out a person from a few feet distance to over a thousand meters distance. And, then there are gun mounted weapons which can be placed in vehicles or fixed positions, and we haven't even talked about explosive grenade launchers, mines, C4, and one which I never hear the passive nature loving noble person mention, the TASER.

Knowing the limitations and uses of weapons will increase your survival even if you don't have that desirable fully automatic .50 Cal M2 machinegun mounted on top of your bunker with remote controls, crates of ammo, panoramic viewing and night vision capability. Chances are you won't have legal access to such weapons, especially weapons like claymore mines, C4, or 25mm miniguns. But you don't have to be a weapons or demolitions expert to improvise or have a plan to get those things when the legal system has been swept away by a super volcano. Of course, the legal system pertains to the US and most other countries, but there are locations in the world where you can live and have those weapons and materials available. Not knowing about critical things like weapons can kill you in the end.

It also depends on what crisis you are facing. If there is an economic or food crisis, the country which you are living in will institute marshal law and the authorities in power like the military will try to keep the peace by enforcing laws. So if for some reason you live near a military installation or near a source of weapons, legal or illegal; be aware that the personnel guarding those weapons will not take kindly to you trying to get weapons you desire to increase your security capabilities. There are also other people you may have to deal with who are also trying to get weapons. They might be friendly, neutral, or hostile. Once again, it pays to know your weapons, and have a weapon with you if you do decide to roam outside of your established survival property.

A group in California has a no weapons/violence policy and actually threw out a member because the member wanted to be able to protect his family with a handgun. The idea of the group was that God would protect them. I do believe in God and know Jesus as my Savior; but don't confuse a stance of non-violence with a stance of trust in God. Many people of God killed many people in the Old Testament, not that it was all right, but yeah, it was in accordance to what God said to do. Jesus came to give us eternal life and freedom, not to free us so we can stand in the face of death to die without lifting a finger. No, Christians are to preach the good news of God's grace. God never said don't defend yourself with a weapon. In the end, weapons are a fact of life, and if you say you don't like weapons because they kill people; then you are believing a lie which many people have come believe. People kill people, whether it is with a knife, hammer, car, firearm, rope, and the list goes on, the firearm or weapon is for your protection not so you can murder someone.

Understand that weapons include yourself. Your body and your mind will help provide security for yourself, family and friends. You should never be limited to firearms alone. If you are in your bunker and there is a person outside with a firearm, they really can't do anything to you unless they have something else that can get to you. It can be explosives, fire, acid, or their own intelligence. If a person doesn't have a firearm, but maybe a knife and you meet them outside, your knowledge and practice of hand to hand combat may be your only weapon that will save you from being stabbed to death or get your throat cut open. Your practice with using objects lying around as weapons might save your life in a physical struggle with another person. If you plan for crowd control outside of your bunker by using CS gas, or Mace/pepper compounds, it will help more if you use your strategy skills to complement those weapons with more lethal weapons like a flamethrower, which by the way, is illegal to make without strict legal approval. But there is the Molotov cocktail, the only problem is making the launching platform, and finding a way of making it automated so you don't have to leave the bunker to use it. Anyways, the idea is to incorporate combat knowledge in your survival checklist which should include firearms, hand to hand combat, improvised early detection systems, explosives if possible, traps, vehicle weapon systems, use of obstacles, surveillance systems; and low and high tech deterrence weapons like skunk or pepper sprays, bee hives, moats, flame weapons, or catapult/sling shot weapons. You might think the later to be weird, but not really if your bunker is on a mountain side and your best defense is to pour hot liquid or rocks down the mountain side. Catapult systems can greatly help in targeting people and objects far away before they get really close to the mountain side.

So what have you learned about those things you see and hear about survival? Your mind is the best weapon, but physics will always win if you ignore it. Some doomsday preppers are sure that they will survive an economic crisis or world moving disaster, but I have yet to see a prepper who even has a ninety or higher percent chance of survival according to experts and logic in the show Doomsday Preppers. Of course it all depends on what you are trying to survive or get through. If you have an economic crisis, then those large families or band of people who have stored food and resources to last several years, who have also prepped with skills and weapons to defend their resources, have medical training and resources; then yes, they have a very high chance of getting through the crisis without being run over by desperate people. But when the prepper is trying to follow what they see others do, then beware of who you follow.

I have seen many preppers spend a lot of money in body armor, weapons, chemical masks, combat training, and much more, but when you analyze their efforts, they don't have the resources to combat a super volcano, the impact effects of a massive meteor, or a dirty bomb. They are not prepared to fight off bacteria or disease brought by dead flesh or contaminated water. They can fight a person, but not contaminated air which goes everywhere outside of an air filtered or self contained atmosphere. They have limitations because of their current occupation, other family member's occupation, lack of money, or understanding of what is needed to survive.

Location, location, location, is one of the big factors in surviving a super volcano for example. If you live in a state near Yellow Stone Park your chances of surviving an eruption is very slim, even if you have an underground bunker with all the bells and whistles as described in this book and other sources. The amount of debris which will fall on top of your bunker will be so great that you might be alive in the bunker through some miracle and your life support system might be perfect, but you will probably live the rest of your life buried alive without a viable plan to dig yourself out.

So the preppers with the best chance of surviving are those that are in the right location for the right reasons and have all the bells and whistles. It is not the person who has bought a readymade bunker that you place in a hole in your backyard and think you have it made. It is not the person who plans out a route through their home city to escape a rioting city full of people going hungry and think you will have a better chance in open ground with everyone else who is also trying to escape the city limits like you. You will all be competing for the same food source or try to take your established safe haven you have stashed away. The preppers who think that living in their small apartment in a dirty bomb situation is going to keep poisonous particles from seeping into their unfiltered air supply, will have a very bad day when things hit the fan.

I'm not saying you shouldn't look at shows like Doomsday Preppers or gather information on survival. On the contrary, you should look at all those shows. You can make your own decisions as to what works, might work, is plain stupid, or should be researched more. What works best is to have a basic idea of what works and what probably will work. I say probably, because none of the full blown preparations have been tested in real combat, sort of speaking.

There are installations like VIVOS, an underground facility established for self containment for a year plus, but even these have issues that I really don't want to go into detail, because it sounds like I'm bashing everything. I sort of am, but the only real and current self-contained environment that I know which comes really close is a nuclear submarine which can survive in the ocean without coming up for air for months if not years without the need to make port for food, air, water or other supplies. The only thing a submarine would have to worry about is the ocean water being so contaminated with toxins that it would not be able to filter the water it takes in, or a massive force like a mass of volcanic debris burying the submarine in the new ocean floor. Otherwise, a nuclear submarine has a very good chance of surviving <u>most</u>, not all, of the doomsday scenarios. Chances are very high that you don't work in a nuclear submarine, don't have the money to buy one, don't have the ability to make one, or be invited inside one just before the catastrophic event happens. But, don't lose hope, because the preppers you see on television or read in books have the right basic idea. We have to prepare and what it takes is you making your own submarine or space lab environment in the ground on land.

There is one other source of ideas, but it also has many misconceptions. It's on the big screen and in writing. In fictitious sources, people see things like "The Stand", "The Omega Man", and "Planet of the Apes." The extreme but in a comical way are movies like "Zombieland". These types of movies depict a catastrophic event which places very few people living in a world full of challenges, which is not a new idea and they try to depict the survivors as triumphant for the most part. However, walking around in a radioactive environment is not something anyone can do without being exposed to the point of death if they don't know where not to walk or stand. The absence of people does not mean people will not be able to use active electrical power sources or all man-made structures will be overrun by wildlife. Without human intervention, there will be things like plants which will destroy and overrun a lot of what man has created. But nature will balance itself. In time, radiation will diminish, granted it will take a very long time in certain areas and concentrations, but the idea is to know what to avoid. Looking at movies of these sorts bring up questions, as to how will civilization continue in reality.

In non-fiction sources, there are scientists that tell you how things will be destroyed or how people will have to adapt.

This is where you can get good information on the effects of global producing disasters like a super volcano, killer meteor, or global warming. Biblical depictions of the last days are usually incorrectly depicted and try to highlight the things which elevate television ratings and not tell the full picture. I will talk about the biblical last days and prophetic view in a simple and basic fashion in a later chapter, and want to place this particular event in a category all by itself, because the end result is unlike all the other events.

Keep your options open to continually do research by watching and reading all sources. Just make sure you view things with a logical set of questions that address how reliable is the information. Seeing people who have experienced the highs and lows of prepping can come from the digital world, and you learn a lot from seeing preppers on video, but it is best to make face to face interaction. So I would start by looking at all that is out there, and decide what you want to trust. Then from there decide what you need to start prepping. Meet preppers if you can, and don't believe everything at face value.

I bring this example because it sort of aggravates me when people talk about Electro Magnetic Pulse (EMP) and how it is so devastating. The Sun is a source of EMP damage as are nuclear explosions in the Earth or atmosphere. But like in TV shows and people who state that an EMP would cripple life as we know it and send us back to the Stone Age is going too far and is simply not true. An EMP will fry electronic components and batteries that are running a circuit; in essence anything that is electrical based and is in use. It is sort of like a lightning bolt frying your TV in a lightning storm, but it will happen to a large area and could possibly be worldwide in a worst case scenario of a solar flare/solar EMP burst. Okay so what is my point? My point is that there are things which are specifically made with insulators to counter EMP effects. Now what; how is that possible? Many new electronics for warfare are double/triple insulated and the circuits are covered with chemicals to keep an EMP from taking down aircraft or disabling a warship. Nuclear power plants, not all, are made to withstand an EMP attack to some degree, by routing electrical systems through a backup system. The control systems that monitor and keep the core from turning into another Chernobyl disaster are now made against an EMP attack. However, an earthquake or other very strong physical attack on a nuclear power plant can cause it to fail and the core will end up releasing deadly radioactive particles into the air. EMPs are not a constant attack on electrical equipment; however, the Sun might emit such EMP for an extended period of time; but that means that a solar flare would have to occur on a constant basis, directed at the Earth. There was a solar flare in July 2013 which missed the Earth and was said to have been the largest ever recorded and could have caused the electrical systems on Earth to bring it to a doomsday event. Chances of flares to continuously be directed at Earth are worse than hitting the winning jackpot lotto every week for a year. So, what I'm saying is that the economy will take a big initial hit in electrical systems, but life is not over and it will recover very quickly. Especially those areas that do have nuclear, solar, hydro or geo power sources. The biggest issue will be the many fires that will be caused when the EMP occurs, other than that, people under estimate their flexibility, and think people are too dependent on the comforts of social order and life.

Note: if you make your safe haven underground, which is an insulation method, if everything is grounded properly, and you have your electrical circuits protected so they don't start a fire, then you will not have a problem with an EMP putting you back in the dark ages. Having said that; a nuclear explosion is temporary, if you have generators, spare batteries, solar cells, wind mills, and the like, you will have electricity for a long time to come. Cars that were off at the time of the explosion may still work, and if they don't all you need to do is get one of your many stored vehicle batteries from your underground safe haven and you should be good to go. You might have to change the starter, but using uncomplicated vehicles (vehicles that are not dependent on a major electrical system/network) will help in them being more reliable. I will talk about vehicles later in the book, this includes boats, cars, doomsday monstrosities, and very mission specific vehicles. The EMP situation will cause major issues in society; but martial law will be placed and order will be established. It is this period where ignorant and fearful people will cause much harm to others. Hospitals will work, and nuclear power plants will also. There are like I said measures in place to withstand an EMP; maybe not in your local area with many electrical issues, but it is not the end of the world or a doomsday scenario. A war brought to your neighborhood for the most part is more of a crisis than an EMP event.

Aside from all that I said about believing what you hear, read, and see; you are the best proof of what is, can, or won't work for you. You can believe the scientist who says EMPs will keep solar cells from working, or you can believe the other scientists who say otherwise. In the end, you need to decide and have backups for everything; which is something all the prepper television shows do have in common. The good preppers are the ones who have backup plans and resources only because they have done their research by looking at all there is out there and deciding what to use in their plans.

Chapter Three

* | * | *

WHAT DO I NEED TO SURVIVE

The premise stays the same when survival is the goal and you are not physically fit, a novice, or without resources. You need to have a determined desire to survive, you need to have a plan which is flexible and is based on sound principles, and lastly you need to obtain resources where ever you can find them. Being physically fit is not exclusive to certain people. Everyone can get into shape (physical and mental) and your desire and determination to get into shape is what you need to start and finish in getting to the desired goals when it comes to your body and mind set. You can be an asset instead of a liability if you have a positive outlook and help those you have found to survive with you. This falls into the cards of life which you were dealt, sort of speaking. You may not be rich or educated, but you can find people who can fill in areas where you are weak in. Your desire to survive and intelligence to adapt is what you need from the start until the day you see your great grand children playing together on green grass.

Resources is another thing you will need. Money, time, access, and support. Money can come in many ways. You can have cash, property, or steady income. I didn't say credit, because it may backfire on you later and what you build might be taken away from you if you die before the doomsday event happens. It won't matter to you because you will be dead, but it might mater to your children or spouse who can't hold on to the property you have built on. You need to consider using credit as the very last resort or not as an option if at all possible. How much money depends on the number of people and location, location, location. A piece of land in the middle of nowhere is probably going to be cheap, but like I said before, if you are wanting to survive a super volcano, meteor, or polar meltdown, then the open flat plains of a continent, especially near the ocean will be expensive because it will probably cost you your life.

So where do you stake your claim of land? Best case scenario: you want a large piece of land in the middle of nowhere at least 1,500 feet above sea level inside a mountain which is not near a fault line, has vehicle access to the base of the mountain, is not facing a nearby ocean, is not a volcano or facing a nearby super volcano, and has a water source above and below you. The Atlas mountains in North Africa facing the south, east, or Mediterranean Sea; Morocco and Algeria fits this criteria. Of course this excludes a majority of the people because getting property, your self-sustaining home and yourself there is very expensive and impractical. The eastern Appalachian Mountains fit most of the criteria except you want to put your home on the western side of the mountain range away from the Atlantic Ocean, and there is a nearby fault line so an earthquake stronger than normal should be a concern when building your home. Israeli mountain region is another good location, the only real constant problem there is living with the chance of a missile falling on your head, but otherwise it is a good location. There are other few areas that may be good, but the following are the things you have to consider in picking a location. If all you are trying to survive is an economic food crisis, then any location that has a reliable water source and is as far away from high concentrations of people is where you want to be. If you are trying to survive a disaster like global flooding, then I suggest you look at water level projections and go live in the areas like northern US, Canada, North Africa, Mid-East, inland Europe and Asia. If you live on an island, I would relocate. If you live in Miami, Florida for example, I would relocate to high northern Florida that is if you plan on living in Florida for at least a generation or more.

If you are trying to survive a super volcano and can't afford to move to the mountain ranges I suggested, then prepare to get creative. This creativeness can cost you a good penny and in the end it "might" not help you, but we must stay positive. I would try to get property that is as high as possible above sea level. Not because a super volcano, meteor, or global warming will make the sea levels go up, but because altitude will be your great ally when it comes to surviving. In the best scenario, your underground home will be in the side of a mountain with your primary water source above you inside the mountain and your secondary source below you. The water which you don't have to pump into your home will also provide you energy like a watermill for example; that's if your solar collectors, and windmills are disabled, or if you can't get a nuclear power plant. VIVOS has food, energy, air, and water, but it is located in Kansas, which is flat. Even though the facility is in a mountain, a super volcano in Yellow Stone will bury the mountain they are in or put tons of debris on the entrances. So if they don't have a plan to excavate themselves out of the facility, then the 5,000 projected population of people there will live and die in the facility, before the younger people can dig themselves out.

The altitude will help you survive from massive waves or rising water levels, and will be your advantage if you are being attacked by hostile forces. It will provide you a means to dig your way out if somehow volcanic or meteor debris does cover your home. Keep in mind that you want to be inside on the side of a mountain up high or at least have access to a horizontal passage way to the outside world. This is where you have to get creative. If you don't position your home in a mountain or hill, then you need a way to be able to dig yourself out if you are buried by earth, water, or both. It is easier to dig down or sideways than upward. Being near the ocean or a very large lake with access to a cliff side may help in these respects. If you position yourself in the great plains of the US, than finding a hill-like elevation might be your best hope, but chances are any debris will bury your home from a few meters of ash or sediments to a hundred meters of solids or water.

There are pros and cons for every build, but the most important thing you need, is to make sure you pick the right location for the right reasons. There are advantages to building on flat ground close to sea level as opposed to taking heavy machinery and building on the side of a mountain at a thousand feet or more altitude above sea level. Building challenges are easier on flat ground, but the reward for building so high will be more of an advantage in the long run. If you don't get the ideal location, there may still be a good chance it will work for you. An underground complex the size of a small town with twenty four people or more might be able to survive in the low lands. It is because they have the resources of a large group and if all of them are on the same sheet of music, they may be able to have an effective way of getting themselves out to the surface after the catastrophic event has abated. I will talk more about numbers and the ideal group makeup, but just so you don't think there is no hope if you can't get to the Atlas mountain ranges, the closest high ground as far as possible from large fault lines or super volcanoes is your saving grace.

After you get a location, you will need a self-sustaining and self-contained environment. The ideal build for your safe home is an underground complex/home. If you do build above ground, I highly recommend you build with features that counter hurricane or tornado type of effects. A dome home for instance is best suited for this purpose, but many people think of a dome shape as just that, dome shape and smooth like a ball. What I am talking about is a geodesic dome home build. It will be easier to custom make your geodesic home with triangular plates in the shapes of pentagons and hexagons, instead of creating a wall that has no flat areas and corners. The home needs to be able to withstand armor piercing rounds from a .50 Caliber and higher if possible. Steel plating more than 3/8 inch thick may do this, but that is not enough. I recommend layers of Kevlar plating, if it is too expensive, then use another medium like layers of concrete and specialized polymers which are available with geodesic homes. There are also mixtures of glass and cement which act like concrete but are lighter and stronger. You can research dome homes and see they are made mostly of wood like most rectangular homes, but you will also see what I am talking about with the triangular shapes and enhancements with different materials. The windows need to be bullet proof and resistant to as much radiation as possible. You need to have all interior wall in your home built in the same layers. You want the home to be completely air and water tight. You need to have sound and water proofing on all the walls. I highly recommend a layer of copper covering the entire shell of your home against certain radiations (mostly from the Sun), and bed liner (used on truck beds) on both sides of the walls or in layers of the walls. On the final outside layer, I would put a layer of the most fire and acid resistant or retardant material I could find and then put a second layer shell with air in the middle, also known as reactive armor plating, to ensure heat or hollow point type weapons are not effective. I have thought about lead plates mixed in with the layers, but that might not be worth the effort. There should be enough material to absorb a good deal of radiation, and if for some reason there is a nuclear explosion, chances are that having your home above ground will only allow the radiation to linger in your structure. The best protector against radiation is hard ground/earth, i.e. underground, the more between you and the radiation source, the better. Aside from that, I would put solar panels embedded with the windows and walls exposed to the sky.

In the ideal build, you would have additional solar panels set up in a clearing on the mountain along with a farm/green house, and wind mills. Note: these structures are on top of the mountain or clearing behind your safe haven, so it is hopefully hidden from people below.

As you can see by all the layers of the walls, and of which I didn't include insulation and reinforcement supports, it would be easier to build underground. You won't have as many problems with making everything bullet proof, or making the walls water and air tight as much as you would with an above ground structure. Your survival depends on you being able to keep radiation and contaminating gases or liquids from getting into your home. Being underground makes this requirement a lot easier (being at a high altitude also helps greatly in this). The second hurdle is being able to sustain yourself in a self-contained area by having good air, water, a waste disposal system, and electricity. These are must haves or you will not be able to survive most of the doomsday events. If you do achieve the safe haven I have been describing, then an economic crisis event will be child's play for you. Depending on your location, someone may never come close to your home for years and you will be able to sustain yourself for many years if not a generation. Having said that, when you build your home, having a low profile will usually help you, like building underground. I won't say always, because if you make an underground home in the North African desert for instance, there is a chance some Nomad will run into your home simply because they roam certain areas that most people wouldn't think about walking through. People escaping a city might accidentally walk over your low profile area too.

You will need equipment for water purification, air purification, waste disposal, room temperature and humidity control, energy storage system, water storage system, (not required but very helpful are devices for taking x-rays, surgery, sonograms), and storage systems which include safe storing of fuels, ammunition, weapons, gases, chemicals, and engineering materials. I will talk more about a vehicle or transportation needs in a later chapter which also ties into the possibility of excavation deeper into the ground or out of a burial situation.

Apart from materials for your home, you need some kind of communications and security system within your safe haven. You should have a radio system setup so you can communicate with the outside world when things hit the fan. If you make relationships with other preppers there is a possibility you might be able to use SAT phones, but that is not as reliable as AM radio, because AM radio waves can travel around the world and does not rely on line of sight or satellites. In addition with this communication system is your security system which is sadly to say very dependent on electricity. There are non-electrical early warning systems, but they are dependent on you being able to hear or see them, requiring you to be exposed to the outside for the most part. There are other nonelectrical systems like binoculars which are passive systems, looking out from a window, but that requires you to be constantly looking through them which is impractical. The security system should be a closed circuit remote control camera system capable of night vision. If you can swing it, get thermo sights. You can get ground surveillance radar systems that detect ground movement and it can tell you a number, direction and distance to where something is moving on the ground. The intent is to be able to see what is going on outside in your safe haven area without you having to go outside or spend all your time in a lookout post. It helps you see what is happening to your structures outside like solar panels, waterfall, and/or wind mills.

You need security is all it boils down to. In this security are your weapons, anti-assault machinery, obstacles, deterrence measures, and counterintelligence measures. I will go into detail about security in a later chapter, but just remember that security is a need, not a would like to have or add-on.

We have been talking about resources, and it comes in the form of other people if you personally don't have all the resources I have said you need. Support may come from people who may or may not want to join you in your endeavor. The support might come from a person willing to help you construct your self-sustaining home, because they are family, but don't really care or believe in all that you do. But since they are family or friends of family they help you in one form or another. There are several prepper communities where you can get support. There are ways of buying old military ICBM (Intercontinental Ballistic Missile) silos which can help you in creating your home, but it can also cause you more problems. Getting an already constructed site might sound good, but it only helps in bypassing starting construction hurdles. If the site is not built to your needs, than you may be spending more time trying to renovate or fix major foundation problems instead of building on top of what is already there. Also keep in mind that installations that are operational or have been placed on standby, like a government nuclear bomb shelter, are very large and require many people to keep it operational. This requires that you have a large group and resources to maintain it the way it is or better. So, my point is you should expect to start from scratch or from a minimal foundation. This will allow you to have a smaller group and not rely on getting very large amounts of resources which only multi millionaires or richer can provide.

The fourth thing you need is a plan and ability to execute the plan. If you are busy working full time and you spend most of your free time going to sports games, bars, watching movies, or even spending a lot of quality time with your family, then you probably don't have the time to be prepping. Time is something that you must use to your advantage. I am not saying you shouldn't have a life, but you do need to prioritize your time and manage your resources wisely. If you are not literally building your home or running drills in your home, then you should be studying on your off time. Studying and learning as much as possible about first aid, medicine, weapons, small unit tactics, demolitions, physics, construction, making or at least fixing filtration systems, generators, communication systems, solar energy systems, hydraulics, waste disposal systems, underground excavation, hydroponics, machinist fabricating, hunting, and much more. Now of course you could not possibly be highly proficient in all of those subjects I just mentioned all by yourself and that is where having a group and a good library comes into play.

Your plan has to start with good research and you must make time for your prepping. You need to be determined, get resources, get a good location, have time and constant access to your location, have or get the knowledge you need, and people to help you. The age of information is here and you need to use both written and digital capabilities for your entire prepping and aftermath. There was a prepper who was showing his two sons how to properly kill a goat and then butcher it. It was done with great care and compassion, which will apply to some scenarios, but not all. There are things that you must have hands on experience with, but regardless, you must have a good database so if there is something you don't know or have not practiced in a long time, you can always go to the 'how to' section of your collected library.

To summarize, there are six needs you must meet in order to survive. The right location, right build of your safe haven, a security/communication system, a well made plan, a good source of knowledge or database, and people to help you. If you analyze the needs, they cover your basic need for food, water, air, shelter, social interaction, and waste management. You will find that many things intertwine with each other and it is a good thing, because when you work on security for instance, you also build group relationships and shelter requirements. The thing people have a major problem in doing is performing the prepping and executing part of the plan without jumping around and wasting time, resources, and effort.

Chapter Four

* | * | *

WORKING WITH A PLAN AND PURPOSE

W e talked about having a plan, but any plan without a purpose is worthless. If you plan to go shopping for groceries once a week, your purpose should be to get enough for one week and get the best prices without sacrificing quality and quantity. If you don't have that kind of purpose, you will shop for things that you don't need, you might go over a budget, or you might not get enough for the week. What happens is that your plan to shop is a plan without a purpose and is considered a poorly thought out plan, or simply said reacting and doing things allowing things to fall into place based on your whims especially if you go shopping while you're hungry. Having a detailed menu for the week helps in having a plan with a purpose and you end up getting the things you need, instead of buying food which spoils or is poor for your health due to your lack of good planning and timing.

Planning for a catastrophic event should start with a simple purpose. Write out the immediate, short term and long term goals for the first stage of your plan. There needs to be a second stage to your plan which is when you get to the point of having a fortified home. The plan should cover short term and long term goals. There are two stages so you don't run into prepping for the wrong things at the wrong time. There also needs to be a break between establishing a home, and where your group are self sufficient and can focus on mastering your survival skills.

The purpose in your plan is to take things step by step, logically and methodically. You can plan to do all the preliminary preparation with your family or partner. In stage two, you can incorporate other preppers. Or, you can put other preppers in your long term stage one plan. This tends to work better, because the people you work with from the start will know you better. A strong relationship is one that will help any plan, even one without a purpose. In time someone in that group will hopefully push you in right direction or purpose. It also allows you to see if the other person is in fact an asset and not a negative influence.

So once again, the purpose of your plan is to have order in how you prep, and what you do once you have the perfect group and perfect safe haven. Within your purpose you have to answer the why and question everything you think is draining the most resources or taking up the most time. That could be changing location because you found a partner and have the resources to relocate to a better spot. It could be you got a promotion or found another prepper who has building materials or knowledge you don't have. You should be flexible and change your plan whenever you can to maximize you getting to your goal as efficiently and quickly as possible. You need to be careful thou because quickness can lead to skipping important steps. Things change and you need to roll with the punches and your purpose will help you do just that. The man who was kicked out of the group in California because he wanted to use his handgun for defense had a plan and purpose. Unfortunately, he was joining an existing group with their own purpose and plan. He wasted time in a sense, but now he knows what to look for if he decides to join another group or even make his own. He should know if the new group members will follow his policies because that will be an area he is paying attention to when people learn about each other.

Your immediate goals should be answering what event(s) you are trying to survive. Once you figure out what you want to survive against you need to focus on gathering information about your enemy. You should make a budget and set aside funds, and other tangible resources to build towards your short term and long term goals. The goals can be very complex, and you should consider that simple goals are not always better. You can be very specific in your goals, and want to be very simple in your actions. Simple procedures are best when it comes to emergency drills, contingencies, and ways of quickly reacting to unknowns. Communicating with others is one example. If you have a hostage situation or enemy are spotted in your security zone, then you must have a simple way of describing them to everyone in the group, and giving orders so that people know what to do without having to decipher code names or special signals. The military has hand and arm signals to communicate information and they use simple number coding or alerts to let the leader and everyone else in the squad know what is going on without having to have a visual on the entire battlefield. I will go into more detail about tactics and responses in another chapter. I just want you to know that your goals should be as detailed as possible, and the way you do things, as simple as possible. You can always change those goal details if things out of your control change or you want to improve the end result. Once again, when it comes to procedures; the simpler the better, especially if you expect young children to be able to follow those procedures.

The short term goals should focus on gathering materials and establishing the foundation for your safe haven. Depending on the events you want to survive, it could be farm land out in the boon docks, or a desert area where people would more than likely not travel through. I didn't mention swamp land type of terrain because those areas are usually very close to or are under sea level. It is also very hard to build an underground or above ground fortification without putting a lot of money into it. Swamp land is an area most people stay away from, just like very large deserts, but in a time of a catastrophic event it might not be the safest or best to survive in. If you are trying to survive an economic collapse then a swamp might be good, except it needs to be deep in the swamps, not next door to another swamp person. Having people near you and wanting resources you have makes for unwanted conflict. Either way, your short term goals should be the foundation of your home up to completion of the structure and surroundings, the on the job training of your group, and continual resource gathering to include information.

The long term goals should be your additions to your home which include added barrier material, extensive surveillance equipment, large scale weapon systems, additional rooms, and backup systems for everything. The goals should include educational and medical considerations. In normal circumstances, education for your children, baring children, and sustaining a healthy group is a must even if you don't have children at the moment. Things like a way to exercise in your self-sustaining home with a treadmill or bike, giving birth, and home schooling. All of these goals you need to write down and talk about. If you notice, these goals are not in the short term goals, but they do need to sort of be there to address considerations in long term goals. Because when you lay your foundation of your home, you need to include the right amount of space with the right amount of rooms. Rooms for sleeping and nursing (a baby, an injured or sick person), a community area like a living room, bathrooms, a kitchen area, a dining area which can be part of an extended work/living room, produce producing room, machinery room, ammunition room, food storage room, a school/library/computer room, recreation room, atmosphere compartments, vehicle/excavation room, auxiliary power and filtration room, a security/intruder compartment, miscellaneous storage rooms, laundry room, and a contingency room. I haven't talked about foundation extensions from the base of your exposed walls to the outside which include areas for vehicles, obstacles, caches, livestock living/shelter areas, and fighting positions.

The second stage plan goals should address improving what you already have in the safe haven. The short term goals of the second stage plan, should be focused on getting more people involved with your group, if you haven't done so already. I am not saying you need to get more people and then start making more rooms to your home. You can if you have the resources, but if you don't have the ideal number of six healthy able adults, three female and three male, then you should get to that point if possible. If you already have the ideal group makeup, the focus should be on getting with other groups to train and improve your skills. I say other preppers because both groups have an invested interest is getting trained and proficient in reacting to hostile enemy or seeing how a good team does things. It could be just teaching your children if other prepping groups are not near you.

Home schooling has become big business for many home schooling companies. However, not everything that says home schooling is good. In addition, novice parents have problems finding or using the right materials. A strong group who can host a week or two at their safe haven or some other location can teach others how it's done. The groups can exchange ideas, techniques, and information which can benefit both groups. You don't want to give trade secrets to any group unless your entire group can trust the other group, so your goals should be as detailed as possible to include what type of people you want to train with and what people you want to live/associate with.

The long term goals of the second stage plan should be focused on what to do once the event(s) have abated to a specific level. The goals once again have to be as detailed as possible. If you planned for a nuclear holocaust or there just happens to be massive radiation due to nuclear power plants going haywire because of a super volcano or an EMP, then your goals must include when and how you should go out into the devastated area. It could be to stay in your safe haven for years or months after your radiation indicators have indicated it is safe. Once the goal of safe exposure is there, the next goal should focus on what to do with your group in and outside of your safe haven. Do you make more babies and get bigger, do you explore and find survivors, or do you go out to be part of the economic comeback? Many things should be asked and your goals should be able to answer many of those questions. If your goal is to make radio contact with other preppers, then that should be in your long term goals if not short term. The fact that you train with other preppers or make contacts will help you achieving some of those aftermath goals. I say preppers, because the preppers will be the ones most likely to be alive or friendly to your cause. You can look for other preppers on prepper chat rooms, forums, and databases online, by word of mouth, and through advertisements. There are pros and cons, but what you want to end up with is to find preppers that will make your ideal group and that are located near you.

I didn't cover all possible goals or plans. You don't have to use this planning suggestion. You might find something better in a remote area of your research, but it is not a matter of finding something that is broken down into ten parts or addresses all the possible issues. No one can cover all possible issues, because you don't even know the possible issues until you start planning and goal setting. Chances are there are people that will give you a ten step process to achieving your goals. But, what you need is a plan with a purpose, not a ten step process on how your goals will be achieved. The plan broken up into two parts before you complete your safe haven basic structure, to afterwards. Goal setting is a continual process which is specific to you and your group. Everyone is different and many goals will be the same in many respects, but there will be goals that no other group will have.

To summarize, the plan you create must have a purpose. To find that purpose you must be able to answer all the questions you come up with respect to what event(s) you are trying to survive, what are you going to do once you do survive, and how do you think you will survive the event(s) to include the aftermath? Most preppers have their family as the foundation for survival. They want to make sure their family is safe. They want to see their children grow and live. It is understandable, but like I said before, if your family is the only group of people that survives on the continent; then you will see your children grow and have no grandchildren unless there is incest which is not the way you want to go. The plan must involve others so you can get to see your great grandchildren, and know that your children will not be alone in their future lives. There are some preppers who believe that many people will survive, like an economical crisis, and group dynamics is not based on procreation or future generations. Whichever way you want to view it, the plan has to make an effort to predict the outcome and see a future with your group, not rely on that many more might or will survive. The possibility of other people surviving in mass numbers is not a crazy idea; who knows, many more people might find this book or decide to prep for the worst and mass numbers of people may survive a global disaster. But are you going to wait for those people to decide to prep, or are you going to start as soon as possible?

Chapter Five

* | * | *

PREPARE WITH OTHERS, NEVER SURVIVE ALONE

The limited or non-existent resources that you might have can and should incorporate others. If you are alone and all you can do is start a collection of weapons, a library, a few food items, and talk to people who are interested or might be interested in prepping for the worst, than the best thing you can do is focus on those people who may join your efforts. If you want to look at it in a different way, maybe you should join a group and help them in their efforts. Resources in the form of money, property, or other tangibles are normally what most other preppers are looking for, but if you have little of those resources, then there is one resource you can market. Your resource is yourself as an able body and mind. Hopefully, it won't just be yourself, and you can include a spouse or sibling. I would try to stay away from a girlfriend or boyfriend, because that kind of relationship can be crushed by mixing with the wrong crowd. I'm not talking about in an intimate way either. What makes a good survival team is the emotional and lasting bond of a relationship like marriage, family blood ties or relations (including adoption), or even lifelong friendships of the same sex. When it comes to the opposite sex, the relationship is either very strong, which most of the times lead to a marriage arrangement, or it is not strong enough to last. I'm not saying that there are no exceptions, but the majority of BF and GF relationships are not strong enough to handle the commitment which comes with prepping for a doomsday event. Reality is that you are prepping for the future, not just an event. A future where BF and GF relationships might thrive is not ideal. However, if there is a BF/GF relationship , I recommend they get with older married couples.

Now the ideal group makeup is six up to twelve adults evenly split in genders. That same group can and should have children, ideally two children per couple with a total of the same amount of boys and girls. If there are elderly, that is okay, but they don't count in the ideal makeup. If the event is a food or economic crisis, then it might not matter if there is an even amount of genders; however, the intent is to have an even spread so relationships are not strained and procreation is easily performed. The reason for having children from the start is threefold. If one person out of the couple is injured to the point of not being able to procreate and they have no children, then all they can do is help out the others to survive, not procreate. If the children are present, they will experience what you experience in surviving and can better manage their children and you as everyone grows older. The last thing is that children are motivators and also strengthen the relationships of the parents and the group.

The number of six adults is important because you need adult supervision at all times. This is not survival by chance, you need people who are hopefully mature, I say this knowing that there are some adults who are not mature or have little compassion or common sense. The couples are examples for the children to follow. If for some reason one or two people are killed, then there will be enough mixing couples to lead and procreate in the future. I'm not going to go into all possible scenarios, but you get the idea; with numbers there is greater chance of the human race continuing. The other reasons are that with six adults you can specialize in areas which will help the group and have better cross training. I mentioned twelve adults earlier, and with twelve adults you can specialize in all sorts of areas.

In the military, an infantry squad is normally a ten to twelve Soldier size element. Each squad is broken down into two teams, A Team and B Team. There is one squad leader and each team has a team leader. In each team you have a machine gunner with an M-249 SAW (Squad Automatic Weapon), a designated gunner (sharpshooter with an M-16 or M4), grenadier (with an M203 Grenade Launcher or M25), and rifleman. Ideally, there will be a forward observer and a medic attached to the squad. Each Soldier has a job in a team, which means that you have two of every job in a squad. There is a primary and alternate in every squad. In a Special Forces team (which is roughly four to six Soldiers), each Soldier is crossed trained to know his own job and the job of another Soldier in the team. One Soldier might know weapons and medical (level of proficiency where he is able to perform most surgery), and another Soldier will know communications and medicine. Each Soldier is also trained and competent in teaching civilians and para-military on their expertise and the art of war. My point is that you can have six subject matter experts among the six adults and they can cross train each other in case one is injured or unavailable. The areas of expertise you want to make sure the group knows are: weapons and small unit tactics, medicine (as close to doctor level as possible, general practice – dental and other areas are a plus), electronics/ computers and communications, engineering (electrical or civil to include mechanics (vehicle and generator knowledge a plus), education (ESE instructor competent a plus), any food expertise (botany a plus), biology or general science expert (this is your air and water purification specialist), linguist is not a requirement but may be needed if your safe haven is outside of your original language base, and general knowledge expert (this is your librarian, or general purpose person). Of course, there are areas that can be more specific to what you should have like a radiologist, certified nurse, and things which many people in society today would not be able to survive normal life without like the technicians that fix hospital equipment or air traffic control radars. The fact that people in the community have knowledge like the later is not a necessity in survival, but in the long run it might help in ways to give the human race a boost in technology that was lost.

There are preppers who are doctors, nurses, police officers, and the like, but if you don't have a group with these professions, you can learn. The areas of expertise are not exclusive to a lifelong career. If you study to be a paramedic or nurse, you will learn enough to be able to perform life saving measures as opposed to just basic CPR and splinting fractures. I'm not saying you have to get certified, but I am saying you have to study or at least someone in your group has to study and become proficient in the area. I spent twenty years in the US Army as an Infantryman and in the military intelligence community, enlisted and as an officer. I am a subject matter expert in special operations, small unit tactics, use of heavy anti-tank weapon systems, handguns, light to heavy weapon systems, and am counterintelligence officer and interrogator qualified. I know security and intelligence gathering methods. I know about numerous Weapons of Mass Destruction (WMD) delivery systems and how to minimize the effects on the battlefield. I know hand to hand combat and close quarter operations. I know all of this due to my training and hands on experience in the military. This expertise would make me an asset in a group, but there are things I didn't learn in the military which also make me an asset. I know about computers from the beginning since they became popular because there were computers available at work. The military forced me to learn computer basics when software is concerned, but the advanced things on hardware and software I learned from other people in the military who I befriended. My friend Brian, taught me how to build my own computer from scratch, up to installing the bios, software, networking, and much more. So, my point is <u>if you know someone who is a subject matter expert, pick their brains.</u> There is also knowledge and skills you have that you might not consider an asset to the group, so you should as a group get together and discuss what everyone can do, knows, or is willing to learn or improve on.

If you don't have a group, and don't have a skill that a group would need, then I would in the early stages of your prepping become a subject matter expert in a subject you like or at the very least think you will do very good in. Let's suppose you have been a homemaker almost all your life and that is all you know. You haven't graduated high school and think you don't have anything to contribute. You might have children, but if you don't its okay because you love kids. In fact you baby sit for many happy parents who are willing to leave their children with you. You might know how to cook, clean, and raise babies. But that is not all you know, because I'm sure you know how to read or tell your kids stories, you might watch TV and know a lot about Oprah, or Jeopardy. May I suggest you figure out if you like teaching kids, nursing, cooking; whatever it is, what you like can be used for your basis as to what you will be hitting the books on. If you like cooking, but aren't good at it, you can try to learn how to cook without the use of a microwave oven and use a conventional oven, or Nu-wave oven. For survival purposes, it will be knowing what to eat, how to prepare it, and how to preserve food which is more important than making the fanciest meal ever. For example: cooking a whole turkey feast in a self contained environment should not be a meal you should be trying to master as a priority. So if you can cover all the necessary nutritional needs of the group; that will be the skill people in a group will embrace with a grin or mean face, but at least they will be alive and nourished.

If you like teaching, then you can learn about being a teacher. You can actually go to school and get certified with a

Bachelors degree and all, but you really don't have to go that far. I would suggest you go volunteer as a teacher's aide and learn as much as you can about teaching. If public schools don't offer that possibility there are many, many private schools who would let you volunteer as an aide (some with pay). The intent is to learn a skill, not just go be an aide so you can work your way to a paying teacher position, or learn how to take care of pre-school children.

I know a woman from Cuba who moved to the US when she was 18 years old, and due to circumstances beyond her control she lost all her high school and college records. She went to a private school to be with her disabled son as a parent and ended up becoming a teacher's aide. Her focus was working with ESE (Exceptional Student Education) students who are students with all kinds of disabilities, because she wanted to be able to care for her son's education better. She ended up teaching at the private school as a teacher for Art, Spanish, and ESE middle and high school core subjects. She had no prior experience and she taught for eight years before she decided to try to get her GED so she could be eligible to get a college degree. She was not a certified teacher, which in a private school you don't have to be certified to teach, but you do need to show competence. She showed them her competence to teach while she was the teacher's aide and grew from there. She got her GED and is currently half way through a BS in Special Education. She is doing her university work online, with a hands-on certification (teacher supervised in the classroom) at a public school for the last six months of her degree. She has been teaching for ten years now and knows how to teach not just your average student from K-12, but your disabled student who has a learning and/or social disorder, or physical impairment. Once she is certified, she can go to any public or private school with not only a degree, but ten years of experience. Now, does this mean you should go the same route?

No, you don't have to get certified or study at a university; however, what you can do is just learn to be a teacher, competent enough to do home schooling, and a little bit beyond that to include subjects that home schooling doesn't go into depth with or cover.

The librarian job is also one of those areas which you can focus on. Knowing where to find information and getting it quickly will help the group in enormous ways. Knowing computers and building a database of information will be something most groups ignore and should not allow to fall by the waist side. People do research and end up putting it in a pile on tables, computers, or notes. Once the person or group knows how to build their safe haven, they throw away the manual and when something breaks and they can't fix it, then they have to look for the blue prints they threw away or tossed in the back of the storage room. The librarian is the keeper of time and history. He/she keeps everyone alert to the timelines and documents everything that the group does and will do. He/she tracks where everything is to the centimeter. Space and usage of that space is very important. The librarian is responsible for knowing where everything is in the safe haven. The librarian can be the accountant to some degree if there is no official accountant keeping the books. Having said that; there should be a separation between the accountant who monitors money, and the librarian who keeps that person honest. So you can have the librarian be the accountant, but you must have someone else in the group who is not blood/spouse related to keep him/her honest.

Another area of expertise which is not thought about very much is the person who knows refrigeration, appliances, heating and cooling systems. You can go to school for a year and get certified in these, which I do recommend you do if you are interested in these areas. Simply because even though you do have to spend time and money on schooling, the end result can help you get your foot in the door and get that experience. You can try to ask someone to teach you, but chances are that won't happen, and you won't get hands on experience without being certified. The good thing is the schooling is not that long, and you will normally have work lined up once you get certified. If you are already certified, I would add to your training and learn about solar, wind, and water energy. That is also something that is hard to find, even at college level. You can work your way into the solar energy business without schooling, but it is very hard to get that training outside of the few schools that provide that training and an apprenticeship type of education. Learning to be a machinist is also a very valuable skill.

If you are not a school loving type of person, and you want to get an area of expertise to add to maybe one you already have; I would get with other preppers or experts that will train you. Unlike being a brain surgeon or computer networker, most of the survival areas of expertise the group needs can be learned from someone who is willing to teach you. You can learn to be a doctor, if there is a willing doctor with the time, and resources to teach you, but that is not something you will pick up in a few weeks or a year. Learning to fire and master firearms is a skill you can learn in several months, depending on how much time and resources you and the instructor can devote. Bottom line, you can learn an area of expertise if you are willing and if there is a means to get you the training, formally or informally. The makeup of the group will greatly help in this because the more people, the more experts there will be that can train you, and vice versa.

The six adult team must have checks and balances, and being an adult is part of that responsibility. It's okay to have a teenager take on some of the adult roles, but it is the overall responsibility for the adults to keep everyone focused and not skip necessary steps in prepping and afterwards. You should look at teenagers as force multipliers and your replacements. They should be your shadows and be allowed to grow but under adult supervision. Once the teenager is competent enough and all the adults agree to the competency of that teenager, then he/she can be included as an adult and have decision/voting rights.

Another check and balance - there is a doomsday community in central Florida which have come together in response to a breakdown in the economic system. The community has a basic town council and mayor. The leader on security is a man that well to make the story short believes in capital punishment in the form of hanging and displaying it in plain sight to deter crimes committed by the community. The rational by this person and I assume most of the members is that law and order must be upheld. I'm not here to approve or disapprove the use of capital punishment, but this is a fact of life they shouldn't want to dictate by one or a few people. The economic system might go haywire, and even if the country goes into chaos, the new government will have the deciding say so on who is put to death. And last I saw Florida does have the death penalty; HOWEVER, there is a due process system which goes to the state and federal Supreme Courts. If the community is going to follow the law to the point they want to, then they should hold the criminal until the new government is back on line. One of the members suggested exile from the community, but this was rejected by the verbiage, you passive liberal. My point in all this is when there is so called law and order; it needs to focus on survival and not example making for deterrence when members of the community or group are those examples. They say they are a community of preppers, but not really. The community should not work to keep order by the same system that broke down, or by the same system but not all of it. Exile will keep everyone from regretting life later. It might not seem like it, but there will be a line drawn when more than one person is executed, and it won't be pretty. It is also a better example for the children. Another issue is the fact that a crime was done; and I'm sure that there won't be DNA evidence to prove a person is guilty or not guilty. Who actually determines guilt? Are the people in the community competent in legal matters, lawyer, judge, forensics, investigations, and so on? Now my point is the moral dilemmas you face will have to include a way out of having to kill someone - supposedly a friendly. Now, having said that. If you are in the ground, and there is no way to get outside, then it is because there is no new government until you get out there and repopulate the race. You are the legal system and if you decide to execute a person, then you better have a strong conviction that the person in fact committed the crime beyond a reasonable doubt, is severe enough for the death penalty, and you have a way to dispose of the body, ie, into ash. Otherwise, I <u>highly recommend exile, or imprisonment</u> if you are in a self contained space until you can get that person outside to perform your supposed justified execution.

There is a chance the exiled person may give away your location to include security gaps and all, but that is something you need to address in the form that if that person does come back he/she will be treated as a hostile and then you can shoot them. What is the difference between the gallows and a defensive kill? Believe it or not, the human conscious will not lose much sleep over a defensive shooting as compared to a slow and deliberate choice to kill another person who you know or knew as a friend or relative. I will talk more about mental therapy later, but in the long run your survival chances will decrease dramatically if you go down the path of executing your own group members.

The group must all have core goals, similar goals, and moral ideologies. They do not need to have the same or similar religious or scientific beliefs. It would be good if they were all on the same sheet of music, but it is not a must have. Having said that, all of the adults must like I said have the same core goals. A core goal would be (as an example) to survive with children and repopulate the planet. A similar goal (not a core goal) would be I will teach my children about Buddhism and the other group/parent pair will focus on teaching their children, Christianity. I use this religion example, because there will probably be a time when even though the adults are comfortable and accepting of the others' beliefs, that may not be the case in time or a crisis. An example is when the children are taught their respective doctrines, what happens if the children turn to the dark side (sort of speaking)? The adults from the start must knowingly go into being in the group for survival that this may happen and accept that the child can choose. Now if the child is being influenced by the other adults, then what? Is blame put on the instigators who are pushing a belief? Well, all of the adults must go in knowing what might happen, which is why it is usually best to group up with people similar to your own core beliefs, and moral reasoning. Now, if I were to assemble a mixed group, I would have a group who are completely scientific in nature, Christians who understand grace, and an agonistic group. Well, that kind of closes the door on many religions and ideologies. The fact is that those three groups for the most part are three completely different forms of ideology, but believe it or not, they will be the most effective is accepting each other, making moral decisions as a collective a lot better than other groups, and raise their children with all the freedom of knowledge about the universe with and without God. I am not going to argue with you if a Scientologist, Baptist who doesn't understand grace, and Hindu type of group makeup will make it through better, or any other type of group <u>mix</u> you care to come up with. The tolerance for many religious and non-religious ideologies is very short and come with many issues which is one reason many people go to war and have prejudices. The group is going to be living together in an enclosed environment for months if not years. Not just waiting, but living and raising kids. If you have a group with the same ideologies, you are good, but if you do not, then I would find a group that does, or if it does have to be a mixed group, get the group makeup that I recommended for best results. The intent is that the goals of the group have to be the same and hopefully the ideologies are also the same.

There are groups who have grown up together and they have no problems living together and raising each other's kids with very different belief systems. That is great, then you should have a good group makeup, but I am talking about those people who search for others and when they find them, they get together because of their expertise and contribution to the new group. They may over look that this Methodist (for example) is not really thinking the same when it comes to a hardcore Baptist who believes in fire and brimstone. It might be easy to live with groups like this when you are prepping and don't have the doomsday event hitting you in the face every day. But when the group gets inside their enclosed safe haven; and has nothing to do except act on their faith and beliefs then that is when you will have major problems in the group. I will talk about fixing this problem before it becomes a problem in a later chapter, section on executing your second stage plan; but the idea is that the best trainer in all situations is to train the way you plan on surviving. In this case you will once your safe haven is self sustaining, spend time together for as long as possible until something breaks. This is known as: train as you fight, and fight as you trained. The group must have good dynamics to be able to fix the problems they will encounter, which includes straining core beliefs.

There are many preppers who have joined with another group, but it turns into an unstable mix of who are the adults and which ones of them are the leaders. In most cases, two groups, usually two couples with children join up. The men, are usually the leaders of their household and it translates to the group as well. The man who has the most resources which includes knowledge is usually the leader of the group. I say usually because there are some very strong and assertive men and women who become leaders. If they become too assertive, the other couple disbands and takes their resources with them. The women should have the same amount of input and assertiveness as the men. All of the adults should be strong leaders. In a group, all must follow the chosen leaders, and never contradict that leader in front of the children, unless it is an immediate life or death situation. If you have a group of two couples, unless they are lifelong friends and trust each other with their lives, you will usually have leadership problems. This is why the ideal number of six to twelve adults is best. If there is a leadership issue, there will be an extra couple who will usually be the peace makers and hopefully the leadership of the leader will become better or the followers will follow better.

There should be two leaders, like the Captain of the ship and the first mate (second in command). It pays to watch Star Trek and similar shows which show how the perfect leader relationship between the first and second in command should be. Everyone else is a follower, but at the same time, they are leaders waiting to fill the spot if the leader is unable to lead. The <u>whole group are leaders for the children</u>. Children are force multipliers, and you don't have time to be getting the parent of one child for him/her to do something another parent instructs. If the group of adults are on the same sheet of music, you won' have children being bossed around by different parents and interrupting each other's orders.

The group leader and the second in command usually will

keep the adults from giving the children conflicting orders. It also helps from children going around trying to get a parent to bend to their own wishes. In fact, the second in command should be responsible in ensuring the intent of the leader is met. An example of this would be the group leader says the storage room needs to be inventoried and stocked by 5 pm the next day. The second in command makes sure he/she tracks who was given the task and assists that person by getting more bodies to help if the task is too great. Or help could be in the form of more or better guidance if an issue arises, and the group leader is busy doing something else. The point is that there has to be a chain of command and people must know and follow commands with the final intent in mind. It is said that the American person can find a way to make things work. The American GI in WWII had the reputation that all he needed was a stick of gum to fix anything that broke. His ingenuity was in part to his ability to take the initiative and understand what the leader wanted to accomplish. The GI would make a way to get the job done instead of stopping to question every procedure and wait for the leader to give his approval. The study of the Art of War by Sun Tzu would help greatly in understanding leadership. There is this documentary on the History channel which goes into great detail on leadership and making sure followers understand the orders and intent of the orders. You can see the video on the internet or download it for your library. It goes into information about tactics, and understanding your strengths and limitations which is good for all of the people in the group to see and analyze.

The leader should know how to lead, and studying things like the Art of War will help greatly in becoming a leader. But if for some reason, you are elected as the leader and you don't have leader experience, I would recommend you and the group see an assortment of leadership videos. Videos that are military focused, not leadership for managers. There are many clips in movies that show good and bad leadership attributes and techniques. Once your group has seen a few dozen videos, you should talk about what was seen and thoughts about how you and everyone else would react or not act. In the end, it is a matter of studying on your own to improve your leader abilities, but most importantly it will be how you lead the group which can best be exercised during group training. If you are a subject matter expert in small unit tactics, for example, you will be the trainer and use your leadership skills by teaching and running the group through a battle drill. Having an observer from outside of your group, who knows how to lead, will help you and the group in improving your leadership skills and the group's understanding of how they should follow.

A good leader has confidence and even if you are not as confident as you feel, the faked confidence you emit has to be seen by the group. If they see the leader full of hope and confident in his/her decisions, then they will follow. Of course, if the orders are careless and harmful to the group, then the followers will be split in their loyalty, by siding with the person they trust the most. You get confidence by knowing your area of expertise, knowing the group, knowing your strengths and weaknesses, and being open minded, but most importantly by deciding and sticking to the decision. If the decision is a wrong one, taking responsibility for the decision and learning from your mistakes will help you grow as a leader.

The military uses what is known as an AAR (After Action Review) which is done to get everyone to analyze what happened in a training session or event in order to improve the training or results on the next time the training session is conducted. They get all the participants (trainers and trainees) together and have a conference about the training event. It's a structured method where the facilitator runs through the events in stages and gets the group to talk about good and bad things about the training and about the things people did or didn't do. You don't have to follow this military AAR format, but you do have to brainstorm as a group on how to improve training or outcomes of normal survival living. You can use a dry erase board or butcher paper, but one of the best visual aide/tool to help in your training for your AAR or brainstorming session is to video tape the training. Being able to see what people in the group do or don't do helps the group see the right way and wrong way of doing things. It is a learning experience for everyone, and videos let you see details and are not subject to memory or emotional limitations.

The group is key in your survival as I have covered, but I want to mention something I only touched on at the beginning of this chapter. If you or your partner aren't able to procreate, I highly recommend you adopt a child or children, if you don't already have two. The intent is to get your children to grow and multiply if you or your partner cannot. You as an adult can contribute to the survival of many people even if you can't procreate.

Your group dynamics must be strong from the start. If it isn't then what could happen is the group creates or establishes a safe haven with every member investing in the structure and equipment. If there is a falling out, it will be because the group didn't address the dynamics of the group and the individuals from the beginning. It must be a commitment to go all the way or walk away with only what you can carry. If a couple decides to leave or is kicked out, then what? Do they take half of the bunker with them because they put half of the money and many hours into the bunker? You don't want to run into this problem which may turn into an ugly legal problem. So, it will be to everyone's advantage that your group understands the pitfalls of breaking up if you are not willing to start with a clear mind set and plan.

The pros and cons. Having numbers is important because they will usually add resources. The con is that they will also add more mouths to feed and lungs to fill. In the ideal group makeup you have six adults and six children. Infants are not force multipliers, in fact they take away an adult from the adult base to attend to the infant. If you have competent teenagers who can attend to the infant, then you have given the adults, their adult back. If you are not really doing anything else but living securely in your safe haven, then infant rearing is not a factor, but I am talking about when the infant is born and a month or two later things hit the fan. The adults will be busy during that critical time period and attending to an infant will drain your manpower pool.

In addition with more mouths to feed you will need more space and supplies. Just to put things into perspective I will run down an inventory of tissue paper. If you figure a six adult and six child group (12 people total); you will roughly use 120 rolls of toilet paper a month, which is 1 roll per person every three days. This takes into consideration that some people use up a lot more paper than others because they don't conserve as they should, and then there is diarrhea, sickness, and other factors which use up paper. Keep in mind that this maybe a little inflated if everyone is conserving, but it is better to have too much than find out you don't have enough. You might also be able to do without so much paper if you have a very large and constant water source, but let's say that is your alternate backup for living more than two years in your safe haven. If the tissue paper is not stored in a cool and dry location with AC and a dehumidifier you will lose all of the paper (altitude helps in countering this issue as well). But, anyways – you have to store 2,160 rolls of tissue paper to have one year supply or 4,320 rolls for a two year supply. If you buy your rolls in pallets you can get good quality packages of 36 rolls for about $17.00 (USD) in Florida. Let's say you get a great deal. A great deal is not getting the cheap paper that is so thin you can see through it, and hard as letter paper. If we conservatively price the 36 rolls for $10.00 (USD) per package, then we get 60 packages or $600.00 (USD) for one year. The goal should be two to three years, even though the experts in Doomsday Preppers state that one to two years of food and supplies is the survival goal. So, for just tissue paper alone, it will cost $1,200.00 (USD) for a two year supply. It is $2,040.00 if you go with the $17.00 amount. Well, things add up and this is just tissue paper, but it is not a budget killer and should be in the budget along with every other small to very large items. What I want to point out is that because of this increase in group size, the amount of space that you need in your safe haven increases. A 36 roll package of tissue paper occupies 13" x 18" x 13" space which is 3,042 cubic inches, 1.76 cubic feet. This equates to about 212 cubic feet for 4,320 rolls of tissue paper, which roughly occupies a space of 6 ft long, 6 ft wide, and 6 feet high (a King size bed area up to six feet tall). So you use up a good portion of a room just for storing tissue paper. This of course is a con, as opposed to 47 cubic feet (4ft x 4ft x 3ft area) for four people and at a cost of less than $454.00 for a two year supply. You can actually story tissue paper just about everywhere, instead of one location in a room, but the point is item take up space.

The money issue is really not an issue when it comes to supplies; the money issue is in technical and expensive items like property, construction and material costs, and equipment to make the home self-sustaining. There are also expenses if vehicles are needed or desired. I will talk more about vehicles and transportation in a later chapter. The issue with a larger group as a con is that more space will be required. This is why more than twelve adults maybe too many for a location that cannot handle that amount, and there is a big chance that the people will have disputes with other members because of the large population. All this might sound like too resource intensive, but it just takes time and good planning from the entire group to get what you need for the six adult and six to twelve children group.

The pros with a very small group is that you can prep easily for a good portion all by yourself at the beginning and have the ability to unite with a larger group later. Otherwise you will be on your own, but that is not necessarily a bad thing. If you have adopted children, procreation without same gene mixing will not be an issue in a worst case scenario. Otherwise, you do have the advantage of being able to make a safe haven that is hidden and protected from any human hostilities because of the small size of the group. If you make the safe haven correctly, you will be able to survive a super volcano and the like. You don't have to worry about leadership issues, because the followers will be dependent on the leader makeup of mom and dad since there are no other options available. Your areas of expertise would have to be absorbed by the two adults and any teenagers, which will put a large strain on the individual. Hopefully, you can look at all the pros and cons, and make a wise decision on who you want to group up with or allow joining you.

There are other things which take up space that a group may develop. If there are babies, hopefully you will have cloth diapers, but there has to be an area for the babies or baby. It might not seem like a lot but things add up. Now let's say a group wants to bring in pets. I highly recommend no pets or animals if you have a self contained underground or above surface hidden facility. If you do have any animals, it should be one to four dogs, maximum. They should also be treated as another person to feed and care, except they are more resource intensive. Medical wise they are more resource intensive, otherwise they are the same when it comes to food; however, you do need to have an area for them too; not next to you all the time. Any other pet is a resource drainer with no real value for your survival. Dogs are great for security alerts, finding people, hunting game, sensing bad food, and knowing when something is wrong using their senses. That is why they do pull their weight; but not the other common pets like birds, cats, and so on. You can try to justify a cat for pest control, but there are easier and better ways for pest control than to care for a cat the entire time.

If you are outside or the crisis is an economic breakdown or just chaos in the land, then I highly recommend dogs, half-breed wolves or German Shepherds, and things like donkeys, horses, a hawk if you can swing one, and things that will contribute to your survival. Livestock are separate, what you are looking for is highly trainable and adaptive animals like Shepherds. You can get chicken and they can warn you of a danger approaching but that is usually too late for comfort. They can provide food, but they are also a source of noise and smell for people that might not normally know you are there, until they smell or hear the rooster or chicken. Same goes for many livestock where they benefit the group through food or even killing snakes, like donkeys, but they all have their drawbacks. The general rule of thumb is to minimize animals to zero if possible, and maximize contributions of whoever is living in your group, human and other species so that security is not compromised and resources are not strained more than they have to. A friend of mine has eight dogs, two donkeys, and a few more animals. He also has the space to take care of them. In addition he has the added problem that two of the eight dogs do not get along with each other because simply there can only be one Alpha dog. So the two dogs are always kept separated, physically, with their own areas or time periods they are outside with the other dogs. If the crisis was an economic breakdown, I do recommend dogs, lots of them if you can manage it. My friend also has six children and his wife is an expert animal trainer, so they can and have managed for many years. So, my point is if you make an underground facility large enough with the right amount of space, resources (food, water, air, and waste disposal system), and have the expertise, then go for it. Otherwise you will notice when you practice to live in your safe haven for a good period of time, pretending there is a deadly environment outside, that the pets are making life hell for you, draining all of your resources or time, or neither. If the pet or pets are not reducing your survival chances, then you should be fine in the long scheme of things even if it is an inconvenience during the practice run.

Chapter Six

* | * | *

SURVIVAL LIVING

The logic of survival is not to see things as only the fittest survive. The fittest includes the one who is most adaptive mentally and (partially) physically. You can be the most perfect physical specimen in the world, but if you are dumb as a rock and walk across a highway full of moving cars without looking both ways, your lack of reasoning or mental adaptively will probably kill you and you will be physically fit no more. I understand that Darwin was talking about the fittest as being the most adaptive, which goes along with what I am stating, but he focused strongly on physical attributes that in many cases made a species more lethal, better hiders, or faster runners, swimmers or flyers. Survival is measured by an attitude to endure the good and especially the bad to worst. Even if you are not mentally strong, if you follow someone who is mentally strong and is providing excellent leadership, that other person's survival attitude might just keep you alive.

The group has to be very adaptive in all areas. The group that constantly talks about every possible situation, what they would do if this happens or that doesn't happen will usually be able to react in a positive and timely manner. Reasoning has much to do with all of this, because while in these group discussions, a person might think they are right but in fact are too stubborn to admit they are wrong or feel it's a waste of time to go down the rabbit hole, then they are not allowing for wisdom to give it's input. The rabbit holes are where you take reasoning and strain the limits of the twilight zone which will be your tool to adapt to any situation. This is how you must live in your safe haven, using your reasoning abilities/ wisdom.

The group, for example, might talk about what would happen if a global killer meteor were to hit on land right on top of Orlando Universal Studios Florida. Your safe haven is on the western slope of the Rocky Mountains not heeding my recommendation to not face the ocean whenever possible. The saving grace is that the meteor effects cause a massive wave of destruction in all directions as far as Arizona before it weakens in strength. But the atmosphere is overwhelmed with debris which will blanket a large portion of North America with the strongest concentration towards you and south of you. The water effects destroy everything near Florida hitting Mexico and everything near north northeastern South America hardest. Well you are good so far, until the impact messes up the Earth's stability to the point of forcing the Yellow Stone super volcano to act up and other fault lines to shift abruptly. The added atmospheric debris from the super volcano along with the meteor will probably bring the coverage of the Earth's atmosphere to block out the Sun or at least half of world if not all for many days if not months or more. Things will settle in time, not as long as many scientists believe because they have no faith in the Earth's ability to stabilize itself. So, since you are on the side of the mountain the fallout of debris will be minimal in your area, unlike the people at the base of the mountain, who will suffer debris on top of them and a tsunami due to the earthquake fault lines that run along the Rocky Mountains. Also due to the earthquakes you will feel in your haven; your windmill power system is disabled along with your solar system panels which are covered by mild debris.

Your group has to strategize every possible scenario. Put reasoning in the mixer and instead of saying well all of that cannot happen that way; think again. Why not? It is simple cause and effect reasoning. Like I said earlier in the book, you need to understand physics and never under estimate it. A very large meteor strike on water is very different than one on a mountain or low ground. The impact will be absorbed and distributed in different ways. A mountain impact depending where it hits could potentially distribute the force into the mountain and the mountain will absorb most of the damage. If it hits on low land, the ground will evenly absorb the impact, but what that means is the force will go deep into the crust and this can create a potential chain of bad events with nearby fault lines, volcanoes, and have unhealthy shifts in tectonic plates. If the meteor hits on water, which is like hitting hard ground, the impact will spread once again evenly, but the medium of the water will allow for the force to extend farther causing major damage to a larger area, in particular, the massive tsunamis the likes we have never seen, which it will produce.

The scenario of an economic disaster might cause massive killing of people and rioting. But what if your sector (safe haven property) is being overrun by two hundred people, many of which are women and children? What do you do? Start shooting at the people with weapons and allow the rest to run through your security area and start trying to knock down your front entrance to your safe home because they are scared, want to live, or angry that you killed a family member. Or better yet, these people have no weapons and you start killing a mob of hungry men, women, children, and infants. Well you put out deterrence measures which end up injuring many people, but you still have fifty people running through your area, not knowing if they are going to keep running or not. You listened to my suggestions and you put your above ground safe haven in the middle of nowhere, and created a very low profile even with the above ground structure. You place a smoke generator and demolition prop so that it looks like your safe haven is a burning ruin and people who for some unknown reason decided to walk towards a desolate area of the country are now approaching a burning area that explodes every now and then. Your group may have argued that giving away your location with black smoke and explosions was not a wise thing to do. However, you as a group decided that it was worth the risk and would be easier to make someone think they were heading into a burning ruin which would reasonably have no food or water since it is burning with sporadic explosions reinforcing the idea that investigating to see what is really there is not a good thing to do. Your group decided that the use of the playing dead/exploding strategy would only be used if a group of over 24 people with weapons were detected going into your area and they have spotted your location. You might think that you have to have a tactician type of smarts to come up with these scenarios, but not really. When it comes to strategy it is nice to have a subject matter expert, but the real strength of the group is that you have more than two minds thinking on how to best move your chess pieces.

How do you use this survival reasoning in your everyday prepping? Well when it comes to construction of your safe haven, your group might come up with simple fixes for specific problems due to location. For example: if you are trying to survive against rioting or hostile people in an economic or terrorist crisis, the group might decide to dig a very large fifty by fifty foot hole, deep enough so that an aircraft would have to be within a few miles distance and at over 20,000 feet to see the solar panels you have hidden from ground level view. You have also arranged the vegetation and ground to surround the area so a person would walk around the area instead of walking through a wall of rocky ground filled with cactus and fake and real rattle snake sounds. You would hopefully consider that shade produced by the hole will attract real snakes and if you have to pull maintenance on the solar panels, you should have a plan to counter the possible infestation of snakes around your panels. If you put up a fence line and say 'Stay out or be shot' sort of warnings, than chances are you have already told people that there is something inside the wire that has not been affected like the rest of civilization. And unfortunately, when people are in a crisis situation and have weapons or something at their disposal, like a HUMMER or even a tank, trespassing into private property is not an issue for them.

Granted, they might change their minds if they are threaten with pointed weapons or shot at, but for the most part, you have advertised that there is some sort of stable source of food and shelter inside the wire. Your group decides to put up signs, Government Property, Unauthorized Personnel will be Shot or Persecuted type of warnings, then you might get better results. However, this kind of deterrence can backfire on you. There might be a military unit which goes into your area thinking they are occupying friendly territory. This might be good if the Soldiers are not idiots and don't try to take things from you without asking. They might be helpful if there are hostile groups in the area. On the negative side, it might be a hostile group of people thinking that if they impersonate real Soldiers, they will get access to a government installation and take it over. Your group's knowledge of your neighbors or locals for at least thirty miles around your safe haven will help you in times like this. Knowing what real military and law enforcement personnel look and act like will also help in things like this. The group brainstorming sessions should include reasonable possibilities like this even though it will be unlikely a platoon size element wearing US Marine Corp uniforms and equipment will be imposters.

I have talked about scenarios dealing directly with crisis events, so I will shift to scheduling of the group. The lifestyle you have in your safe haven will depend on how your group decides to adapt and live. For example: your group should use survival reasoning to discuss and plan on how the children will be home schooled, which teachers will take turns to teach, when group teaching methods will be conducted and subjects. The group should know what everyone is doing, the subjects they are learning, and who is running the sessions. An example of this is: The group brainstorms and reasons that if they get two adults to teach the younger kids on basic subjects in the morning, the other adults can take the older children to fire weapons and learn estimating distance in the morning. The adults decide not to fire weapons in the afternoon during the hottest time of the day, and use that time to teach all of the children about what to do if power goes out or water is cut off. This might sound like it is a no brainer, but the reasoning comes in the form of logical use of time and resources keeping the group from wasting time, or training outside in the middle of the heat or cold weather when there is no benefit. I have seen people train at crazy hours and unlike the US military with a lot of resources who can afford to train whenever they like, they do have times when training is conducted only in certain hours of the day to minimize possible heat or cold weather injuries which prevents from wasting resources.

I mentioned that the children are force multipliers. They are force multipliers because the can not only augment your manpower needs, but they are very adaptive and learn very quickly. They will grow in maturity quickly with the right guidance and motivation. There was an episode on Doomsday Preppers were two couples joined forces, and each of them had one daughter. The older teenager was 12 years old and the other was 11. The 12 year old was taught from an early age the life of prepping and knew how to use weapons and understood concepts like not putting down your weapon in a hostage situation. The other teenager wasn't taught from an early age and when her father took her to an indoor range to learn how to shoot a .22 pistol, she was very passive and not very willing. The camera person took separate shots of the girl talking to the audience. She indicated she would probably never fire a pistol again, even though she did shoot very well hitting the targets center mass with a tight shot group. The father said that he was proud of his daughter and saw that she enjoyed it; would get much better with more practice; and would be able to cover his back. Of course the contradiction of their feelings or observations of the training event is not the point here. Being passive is also not an indicator of being immature; but the concept of how important it is to know how to use a weapon was not fully developed in the teenager because she had not been given guidance early on and was the type to be inpatient, easily bored and frustrated. The 12 year old teenager was highly proficient with her weapons and could snipe at a very good distance away. She understood covering a person and not walking in the line of fire or letting someone else get in her line of fire. She was not afraid of the weapons she carried and her feelings about protecting her family which included two younger brothers was clearly developed in her upbringing and reasoning.

The children must have more than a reason to follow you because you are the leader, parent, or adult. When you have a child that has not fully understood the concept of survival, then the group needs to address it. You might think that adult intensive training would do the trick, and you would be wrong. The best teachers of children who are not mature or even rebellious are the mature children who know why they are training for war, sort of speaking. Habits are taught by example, and children usually demonstrate and transfer good and bad habits to other children. I'm not saying you should let a teenager teach the other one how to fire a firearm with blanks or live ammo, just the two of them alone in the wilderness. The knowledgeable teenager can be the instructor and an adult can be the safety. They (the instructor teenager) can walk the children through an operation and explain why they are doing it. They can be the facilitators and you can be the observer controllers. Now you can't and shouldn't do that all of the time, but the fact that a peer can do something and can answer the child's questions and fears is something that will mature the child into what is important and not focus on going through the motions because dad and mom want me to. Of course the relationships between the children will affect this, in a very negative way if the teenager who is instructing is not liked or too bossy. If the teenager who is being trained is rebellious, then that is a parenting issue that needs to be fixed before you can get that child or teenager to see the light of reality. I will talk later in detail about prepping and living with your children and other's children. The child you raise will in time become an adult and if all you trained him/her to know and believe is that a doomsday event will happen and their sole purpose in life is to survive it, then you have failed from the start.

I was impressed to see the 12 year old girl talk about how bad she felt when her younger brothers turned over their weapons to the hostage taker, who had their mother at gun point. She understood that shooting at the hostage taker was the best action even with the risk of hurting or killing the mother. Each situation is different, and I point his one out, because the younger brothers said they didn't know what to do, so they followed the commands of the hostage taker and not the mother or the sister who were yelling out commands, above the hostage takers voice. The simple reality of mom in danger overrode their reasoning for survival. The sister knew and had compassion for her mother's situation, but she focused on not just her training, but her reasoning that giving up your weapons left you at the mercy of the people who are trying to do you harm, intentionally or inadvertently was due to her maturity and understanding of the big picture.

All of the children seemed to understand that a doomsday event is supposed to or could occur which is why mom and dad are prepping. But that is all they really know except for those mature children who have asked their parents more than are we there yet. The children can and should be allowed to contribute ideas. The amount of information and reasons the adults give the children will turn out to be a blessing for the group. You will see that children tend to say what is on their mind and come up with many things outside of the box. In many ways they ask the questions like why are we doing this and doing that. Why can't we go outside, and if we can't go outside what can we do inside? This might sound sort of trivial, but it really isn't. The group needs to be able to explain in detail why they are doing this and that and come up with good solutions; otherwise the child will figure out that you are just making up answers to keep them quiet. The things that can be done inside, for example, should include activities that are <u>fun</u> not just educational. No matter your age or maturity level, you must have fun quality time on a daily basis. If it is listening to music on earphones, playing board games, trivia, role playing games, hide or seek, watching a movie, computer games, that is something that your reasoning should not ignore. Being in an enclosed environment is not going to be good if you have a daily schedule of school work or instructions for the children all day long and a few hours of free time in the evenings.

I vividly remember pulling 16 to 18 hour shifts working in a Tactical Operations Center during military training exercises. The work was tedious and there was time to relax, like during some meal times, but when you do this for a few weeks, you find out that since you really don't have any free fun time, you get stressed out and by the end of the exercise everyone is trying very hard not to lash out at co-workers. Your time management is something that must be conducted with children input or considerations. You don't ask the children out right what do you think we should do all day? No, you figure out the basic schedule with days off from school, training, and work; and make time frames for play, group quality time and allow for individual time. Then you ask the child what he/she would like to do during play time periods. Of course you must consider the age of the child and get input according to their level of understanding. Your reasoning of how and why you do things will be partly based on the likes and dislikes of all the members of the group. I personally could spend months on end just listening to music, writing, and watching movies or shows to be happy. There are others who hate being inside and love the outdoors and actively doing physical things. I love the outdoors and exercising too, but it's not just a matter of one individual or a portion of the group liking something. It would stand to reason that the group does what the majority of the group likes. Playing UNO for instance is a game the entire group can play, but if there is one or two people who don't want to play UNO or similar games once a week, then your reasoning should be focused on getting the mature people to maybe sacrifice a time period of UNO playing to something completely different which the individuals might participate in. Not that you have to do something as a group all the time or once a week, but change is a very good thing when it comes to living in an enclosed area with faces that don't change.

Make sure you celebrate birthdays, and special occasions, even if you think it is not necessary, like your own birthday because you are over 40 and don't really care too much about the event. The celebration is not just for you, it is for the entire group. It doesn't have to be something grand, but at least take photos, sing happy birthday and eat some cake or other deserts, if eggs are not available. Celebrate graduations and make sure they are grand and equal for all the children who graduate, at different levels, not just a grade, dividing elementary, middle, and high school. If the doomsday event has not occurred yet, it could be a college graduation for a degree received through the internet. It is kind of ironic that I am telling you this because I dropped out of school with a GED when I was sixteen in order to go to college. So, I skipped high school without a graduation ceremony and finished my bachelor's degree after having a break in military service without going to a university graduation ceremony for my BA. I did however, see and participate in many military and civilian ceremonies where I and other people received recognition for the achievements and hard work put into school, work, or a hobby. Recognition for something should be something everyone in the group needs. I say again, Needs. If you don't give that recognition even in the form of thank you or well done, the dynamics of the group will start to resent each other. The adults might not be affected too much by a lack of recognition, but the children will see this as favoritism or think something is wrong with them or someone else. Father's and Mother's day should be an event like Easter is treated. The people are recognized and a small game is done so that the children can participate in instead of just making a small gift or something for mom or dad. The great thing about celebrations is that everyone in the group benefits.

This brings up a question on the area of expertise known as Psychology. It would be nice to have a professional psychologist or even better, a trauma crisis counselor in the group. But, chances are that you won't have an experienced psychologist, so your group should do research on behavior of children and trauma due to things like combat, and life changing events like losing a limb, death, or birth of children. Many professions like a certified teacher who is taught on child behavior and teaching practices, are exposed to skills which help in child behavior and adult behavior. Police officers and military interrogators are taught about adult behavior and how to manipulate good and bad responses. A psychologist is considered to be the master in all of this, but that is not true. I have a minor in Psychology, interrogator experience, and counterintelligence training on getting information from people (spy stuff). I have studied on and experience with teaching students and adults, and know how to teach. Am I qualified to tell you how people act and what they need to improve their behavior or learning curve? Yes, but you know who is best suited to tell you what to do to improve behavior, morale, and interest in a subject of life? The group as a collective, the group who are hopefully partial in nature, have compassion for all the children, and have personal experiences in what helped them get through their early childhood, teenage years, and life in general. What a psychologist can or does do for the most part is point out to you the behavior and habits which you or others have taught your children. Children don't come up with behaviors due to natural selection or the influence of the force or dark side. The parents have a major influence, but remember that influence is subjective and changes with time. Visuals like movies, other people, written word, even actions of everyday living, gives the child a basis on how they should behave and also what they should believe. You don't need a psychologist to help you with behavioral problems, you need a collective group who understands and helps without condemning or enforcing unnecessary ultimatums.

Having said that; there are behaviors that do need to be taught and enforced. Everyone should be safety conscious and educated in all areas of possible lethal or disabling injury. A four year old or older should know not to touch electrical sockets or grab pots on the table or stove. Safety should be a highlight in all training events or teaching sessions. Written instructions should be available or physically on all equipment, in particular hand tools, electrical appliances, generators, etc. The librarians should get the older children to label things with instructions on how to use or at least put safety labels on equipment and structures. An example of safety labels for structures are: low ceiling or flush toilet after use. Wash hands before leaving bathroom, or wash hands before eating. I mention these labels, because hygiene should be considered a safety practice and not just a cleanliness or courtesy type of behavior. How you take care of sickness, preventive measures for everything, and dealing with routine medical exams will be something that all of the people in the group need to understand and assist in. If at all possible you want to have two experts in medicine, one male and one female. This will eliminate notions of female and male, or male and female issues when it comes to exposing body parts and talking about possible embarrassing personal issues.

The lifestyle of privacy should be a strong concern when creating the foundation of your safe haven, but also when living in an enclosed space. Sound proofing is a must in your safe haven for many reasons when it comes to getting sound sleep, private time for couples, and minimizing sound in case (hopefully never) firearms are used inside your safe haven. There should be a separate bathroom for each adult couple and at the very least one bathroom for two children, and two half baths on separate ends of the complex at a minimum. Having said that, waterless urinals should be used and out of the six bathrooms, at least two should have a bathtub. I don't recommend you have six bathtubs. It will be a drain on water and it can be easy to have accidents in a bathtub, especially when you are dealing with young children or the elderly. You want a bathtub because you do need to give baths to babies and that is the best location, not the sink in the kitchen. You don't have to use the bathtub itself, but you should put the baby tub in the adult tub. You might need to use the tub to soak someone with Epson salt, ice, or for therapy. You can use them to store water if something goes wrong with your water purification system or waste system. There are probably several other uses for the tub that I didn't mention and it's better to have two instead of none and later be sorry you don't have one when you need one.

The materials which you use for the floors, bathrooms, bedrooms, and storage rooms should be decided on the basis of the budget, but also on quality. The appearance should be the last thing you consider if at all. You are not trying to decorate your safe haven so that it will win the HGTV home awards. Carpet is good when you wake up and don't have to worry about stepping on a cold floor, but it is not hygiene friendly. Wood and very durable tile is good, but what you want to look at is the function of the rooms, and the durability of the materials if they get wet, pounded on, or constantly tracked on by feet, furniture, and shoes. What you want is a floor that has as little grout as possible, durable, and does not cause a safety hazard when wet, or if socks are used. Carpet is good in some cases, but I highly recommend you stay away from carpet. Cork is another good material for flooring, but not in all areas. Your fuel, food, and rooms that have the potential of being hazardous to your health should be floored properly with epoxy, metal, concrete, or special rubbers (usually used in gyms).

Asian culture and people have a habit of living in the homes without footwear. The shoes are placed at the door entrance, usually in a designated area or closet. This tracks less dirt into the house and keeps the house very clean and makes it easier to clean when the times comes (usually the only thing that is cleaned is the dust that collects on the floor). If people keep their feet washed on a regular basis, you won't have athlete's feet from spreading or fowl smells in the home. Now, I recommend the shoes not be used in the bedrooms and bathrooms. The other areas I recommend shoes that are specific to the rooms. If boots are worn, then it is for the machinery room, storage rooms, or in preparation to go outside. Otherwise there are comfortable sandals or running shoes like the ones used in fitness centers or by joggers that do not have laces which will work best in the kitchen, living room, etc; and if an emergency comes up, the people can run outside with the running shoes. What you don't want is boots, hard sole shoes, or dress shoes being used and tracking around the complex. If the group is not accustomed to these practices, then when you have a baby, you will find that germs will spread to the baby faster because of the use of footwear that has a tendency to go outside or in areas where exposure to oil or fuel leaks are possible, like in the purification rooms or vehicle room (garage version of a room).

One last thing on survival living. The database which your group has compiled will be your knowledge source for facts, and entertainment. There should be no pornography whatsoever in the database. Some couples will argue that pornography is only for them and it enhances their sexual activity and pleasure. You will be playing with fire, because no matter your sexual or private intentions, the children will in time be exposed to it and it will destroy any sense of morals you will try to instill in them. Bad habits are bad habits. Smoking is one like pornography which needs to stop. You might say that you will stop when the big one hits the fan. Sorry to disappoint you, but you are looking for trouble if you rely on a crisis to abolish your smoking habit or any other bad habit. Usually a bad habit is replaced by another, usually another bad habit. Sudden or tempered nonsmoking usually turns into changed eating habits or coffee/caffeine drinking. Get rid of the habits which cause health problems and use up resources. I smoked for over 20 years, stopped cold turkey twice, once for two years and once for one year. I tried patches, and time interval reduction methods. I highly recommend you go talk to your primary doctor and see if he/she can prescribe you Chantix. It is a pill you take once a day for 90 days. I tried it for one month and stopped. It is expensive if you think smoking two packs a day is expensive. If you stop and not complete the 90 days it will not work for you. I skipped two months but started back up finishing the other 60 days of pills. You will spend no more than $450.00 without coupon discounts, but I guarantee you will stop smoking. You can smoke while you are taking the medicine, and simply put, you will just not want to smoke after the first three weeks, but you must continue taking the pills for the full 90 days. The medication builds up in your body and the 90 days will make it so your body doesn't crave the cigarettes ever. I have been smoke free for over four years now and it feels exactly like I used to feel about the smoke smell when I was in middle school, feeling it was repulsive. There is the new vapor cigarette, but honestly, it is a wasted resource to have a vapor cigarette in your safe haven and teach the kids it is okay to play with nicotine and believe it's a cool habit. The problem with addictions is they are always there, maybe not for you, but maybe it is for others. Alcohol in the form of hard liquor should not be thrown out of the inventory list for the safe haven. Some items will be useful even though they do present a source for a bad habit. There are certain alcohols which may help you in sterilizing tools, and large objects. You might be able to use bleach, but that may give off more fumes that can harm young lungs and it is not ideal for surgical tools. I would recommend wine be allowed in the safe haven, but strictly controlled and stored. I would make sure all of the group adults and teenagers get professional help now on issues like alcoholism, drugs, mental and physical abuse, smoking, pornography, and a few more before the addiction or habit is brought into the safe haven.

In the end, the children in the group will be the motivators for living life to the fullest and as happy and carefree as possible. Having bad habits and addictions strangle this motivation. Having strict orders with no compassion is also going to strangle the growth of the children and how they will follow you. The calmness and respect the adults demonstrate will allow the children to follow and not be selfish or spoiled.

The adults should be comfortable in their life to be happy and be sexually active with their lifelong partner on a weekly basis if possible, just like the children would like to be able to play every day in the park with their friends. The daily activity in the enclosed safe haven has to be as close as possible to normal activities on the surface. Socially, the best group makeup is twelve adults and twelve children. There is enough social diversity that children will never get bored living with other kids; and the couples will have more time to enjoy each other and the family life with their children and friends, because the adults can spread the wealth of jobs, work, supervision, and play.

There was a prepper who had an underground bunker made of metal up high in the mountain next to a road where there was constant snow. There were two adults and three small children. The bunker was impressive with three years of living supplies and radio comms to the outside. But it was sort of cramp with bunk beds in their living room/bedroom/family room connected to an open extension of a kitchen. Three years is really not that long if you have the supplies and enough space; however, they had barely enough space. You couldn't run around or really do things without privacy or without waking people up or disturbing them. It can be done, but it would be uncomfortable and very stressful. The family was a family and it would be easier for them in a small space, but had it been two separate families and more adults trying to enforce respectful behavior in a very enclosed space, many social problems would have arisen. The routines and things they do as an individual or group should not be completely dictated by a lack of space or taking away many activities that could be performed with a plan in place to perform in challenging spaces.

I mentioned earlier about law and order. The larger the size of your group, the larger or more law and order people and you will need or demand. This is not true. The laws and rules that apply to one person, apply to all. Your security plan or force is not defined by how you can rule over your group. The community in central Florida is a community just like that in any community that proclaims they are separate from the rest of society. In many respects they are like the Branch Davidians in Waco Texas. I'm not saying they are fanatics or socially misguided people; but what I am saying is the power struggles and view of a community is similar, like other communities. So, what is my point? The social environment you surround yourself with should not be based on a social order of a couple of people over many people. If you look at the dynamics of the ideal group you have six adults up to twelve. The leaders are two, one leader and one second in command. In a group of six, that is 1 to 3 or 2:6 ratio of the adult population who are the primary leaders. Anything beyond that requires a better check and balance. In the community of 25 to 50 which is the one in central Florida, there should be a breakup in the group with all the couples acting as vote casters for decisions that are not time sensitive or directly linked to command decisions. In the example prior about capital punishment, there should have been a vote among all the members and how it should be handled. Instead it was forced on the community by the established leaders of two to three people.

There seemed to be no check and balance on the leadership. For a group of 25 or more, there should be a council of five people to decide on command decisions like punishment for minor crimes, where to place resources, and who to train, promote, or demote. Large things like capital crimes should be put to a vote to establish the procedures to include changing laws or adding or kicking people. The group as a whole is deciding the fate of the community. Unlike Waco, Texas, the fate of the community was decided by one leader and his enforcers. Granted there were people who followed the leader, but that is not a check and balance. The power of one person should not extend farther than six to twelve people. Anything beyond that is asking for social disorder if there are no more leaders as part of the final decisions. Note; you want an odd number like a three or five leader council so there is no stalemate.

Okay, my point is if you have a six to twelve adult group, your group is run independently as a group because the few people should keep the leader in check when it comes to social order and laws. You might think that if there were more people in a group like 15 or more than the people would keep a leader in check but that is not reality. A leader can say something but if there is only two or three people who object, then the leader continues, even though not everyone might agree, they just don't know any better, are scared of the leader or don't want to decide and allow the leader to decide for them. In a small group the leader needs to listen to the members; otherwise he/she loses members or the leadership position physically or socially (like losing respect). You can link with other groups, and help each other out and live in close proximity, but when you proclaim the groups as a community you are not one small group anymore and you are at the mercy of whoever becomes leader or group of leaders, if you don't have voter input or some sort of group influence.

The movie, "The Stand" is a good example of how there was a council to lead the group on the good side (with also a primary leader), and on the side of evil there was one evil demon in charge. Not that Hollywood is not dramatic to extremes; but yes they are. The show, "Falling Skies", also has a demographic of leaders and loners/small groups, but they showed the social order side by trying to mix in law and order by using a military style type of order and discipline. Leadership in the form of a chain-of-command is present and they try to show conflicts with social order and leader challenges, but it is a show. In real life, people will try to do the same. So for the best ideal survival group I highly recommend groups of six to a maximum of twelve adults. In the movie, "The Stand" the good side had a council of leaders which the people were able to rally behind because the community as a whole knew that decisions were being made by more than one person. This allows for a leader to be wrong or just not with it that day because the other leaders can take up the slack of good decision making.

Now, I need to talk about people on the move. Depending on the event you are trying to survive you may not need to worry about anyone coming to hurt you or take your things. However, in the long range scheme of things, you will hopefully run into people who are friendly, even if the world is covered by debris and the only survivors are other preppers coming out of the woodwork. In a crisis that allows for many people to be out and about, you can end up with major problems. If you decide and I highly do not recommend this; but if you decide to make a group and go help people or in a negative way go take possession of what is not yours; then here it is: You decide to create a squad or platoon to find people and help them, but you must also protect yourself from terrorist or bad groups, then you must work as a military unit; using military methods, not police or hunting party methods.

There are heavy mobile doomsday vehicles, which are made for families and for fighting. The best for your purpose is a heavy wheeled armored personnel carrier, or an armored HUMMER. Light to Heavy Tanks use up too much fuel and have unwanted logistical problems you don't want to deal with. Your weapons of choice should be vehicle mounted high caliber weapons like an M2, or Mark-19. Now, how are you going to get this? Well that is best case scenario if you come up on military weapons and vehicles you can salvage or obtain. What you can reasonably get are .50 Cal weapons, mini-guns, modified assault rifles, homemade mortars, and launch systems like homemade missile launchers. Many people think fully automatic weapons are superior to all the other weapons, which is only true if all you're trying to do is put lead down range to suppress fire. Since you are not in the military, going fully automatic is illegal in a few states and a waste of ammunition. Fully automatic weapons are above all inaccurate compared to semi-auto. Except for those weapons that are bipod or tripod stabilized. But, since you are not in the military and do not have a resupply of ammunition, that will be your deciding factor in carrying or towing an ammunition eater.

Your safe haven needs to be able to withstand against what people can carry, have access to, or tow around. Your group that is moving around has to have access to food, water, shelter, and ammunition. Usually this comes in the form of cache points or expected sites which have fuel or things you need, like a hospital or another safe haven from a friendly prepper. Whatever the case, no matter what is going on, your group is always vulnerable to the environment, people, wildlife, and problems of your own making. The rule of thumb in order not to create a self inflicted problem is simple. If other preppers have prepped properly and are hostile or unknown, then when you come up on a location that seems fortified or well protected, leave it the hell alone and find easy targets or locations to resupply your needs.

Doomsday Castle is a good example of this. Not that the castle is not well protected or impossible to conquer, but simply put; it will take resources to take over the castle. Not because it is built with thick stone, is tall, is on a hill overlooking down in all directions, is occupied by people with weapons; no, its because you will need to spend time and effort to snipe people, get a weapon to launch a fire damaging arsenal to smoke them out, fight up hill, or launch attacks from adjacent hills, and use all of your group to do it. And what do you really end up with? A few if any survivors in the castle who hate you and will be a problem for you, dead bodies, a cache of supplies and materials maybe. But you will probably have lost a member of the group, and spent a lot of effort, when it would have been easier to storm a warehouse, farm, or construction site and make your own castle or bunker. It is simple, avoid hard targets. If the people around your safe haven are just as smart, they will also avoid you and pillage and plunder easier targets. But not everyone thinks this way, which is why it's best to have a strong lethal response so they do get smart quickly and move elsewhere.

The main message is you should not be moving outside unless the crisis is over and it's time to rebuild, or you have to in order to restock your one or two year supply. My policy on stocking supplies is you should have five years worth which can be replenished within your property. I must also address that an economic collapse is a doomsday event, but not really. The stocking of supplies is commendable for such an event; however, there are many people who stock up for this event but very few have a way to protect their assets. Some preppers have an impression that a few guns or no weapons are going to keep everyone at bay if people are in despair and will even go as far as to destroy your stockade just so no one can have it.

My point is you need to be in the position to withstand large numbers of attacks, fire, smoke, acid, chemical attacks, armor piercing weapons, and biological attacks; hence, the underground fortified with lethal weapon measures safe haven is best suited for protecting your group, supplies, and food. You will have to determine how you will operate outside of your safe area, and if you go out there alone or without the intent of matching force with bigger force, then you are vulnerable to being an easy target. Even if you do have massive firepower and a lot of people with you, if you project to be an easy target, you will be. An example of this is a platoon size group of snipers who are by nature suppose to be invisible, but let's say one person is caught out in the open or is discovered. Well to anyone, logic would suggest that one or two people are easy targets; however, there are 34 other people with high powered rifles, grenade launchers, and close quarter weapons who are invisible and have the element of total surprise.

If for some reason you decide to move around with a small force and think that making your small force seem like a very hard target, like driving around in a tank, then I highly recommend you look at all your pros and cons. A tank is a very hard target; however, what makes it a very hard target is not its armor, weapons, or movement capabilities, it is the fact that all armored units have support, in particular, dismounted support. It doesn't take a gunship, antitank mines, or specialized weapons to destroy a tank in the end. It is easier with these types of weapons, but one thing tankers fear the most in an enclosed area like a city or town environment is dismounted people capable of throwing Molotov cocktails or incendiary grenades on top of the tank. IEDs are also something that is not easily done or placed in an unknown area, but it can be done stealth fully with people on the ground. Having said all of this, your best weapon against any enemy is stealth and not allowing the enemy to know what you have or don't have. Once you are in a fire fight, it is still an advantage for the enemy to not know where they are being attacked from, how much firepower is not being used against them, or how many people they are dealing with. If you have seen Mel Gibson in "The Patriot" when he attacked the British Soldiers, trying to save his son, you will understand that his advantage of surprise and confusion is what helped him get a victory in the middle of combat. Your ability to hide and stay hidden will greatly increase your small group's chance of survival. If you have a large force, 40 or more people, that you have grouped up with in a post crisis situation; it is still good practice for the leaders to make sure the group's combat power is concealed and they do not demonstrate a show of force. The only time you would want to have a show of force would be when you have an army which is basically over 8,000 people, equivalent to two US combat brigades.

Just because I do try to take you down rabbit holes, you don't have to agree or learn anything out of it. Stick to the level of where you are, and if you have a small group, anything with less than 12 people, I would try <u>not to make</u> my location or activities known. The rule of thumb is the larger your group, the more options you have when it comes to moving around on enemy or even friendly territory. Vehicles make this stealth advantage almost impossible, which is why I suggested mountain bikes or horses/donkeys to move around if you really have to. If you know for a fact, your gut feeling, that there are no large hostile forces out there due to the waste land of a meteor for example, then by all means don't worry too much about being too quiet and take vehicles, large and small to move around. Aircraft would also be very helpful and a strong advantage in combat and other activities.

Chapter Seven

* | * | *

A WELL ROUNDED PLAN IS A GREAT PLAN

T he plan with a purpose is good, but unfortunately I need to add that it has to be well rounded to be a great plan. It has to cover all the major factors from start to finish and the unknown. We have covered the base plan which is specific to your needs of the location, what you are trying to get through, and the group makeup. What I am addressing here is the need for a full range of contingency plans. You have to be as detailed as possible in your goals like I said earlier. This will help you develop contingency plans that cover just about everything to include unknowns where something happens and a person thinks out of the box and finds a solution using an adaptive form of a contingency plan.

If you don't plan on storing a foldable wheel chair you will end up having to keep an injured or sick person lying down or in a fixed chair. This is not good because it takes another person to care for the injured person. There may be no infants or toddlers in the group, but if you don't plan on babies in the future you might end up with a baby and have to create makeshift clothes, baby friendly medical items, and alternate food items. It would help greatly if you plan to store baby and toddler supplies. If you think that people won't get sick from minor sickness to major sickness, then you will run into problems when you don't have the means to quarantine a person, or worst case scenario not have the anti-venoms for snakes in your area. Your medical expert should ensure that there is a plan for medical needs. Simple things like everyone knowing everyone's blood types, allergies, medical history, and having stored plasma and general medicine for infections. As you can guess, the adults are given tasks in the areas of expertise, and make draft contingency plans to cover their areas with all possible scenarios. The group gets together and each contingency plan is analyzed and perfected or completely changed.

The plans that cover everything are created as a group. If for some reason you are a small group, then my recommendation is for you get with other people, not necessarily other preppers and discuss the contingency plans you have come up with. Depending on who you asked to help you, you might get responses like "you are crazy" as they hear your plans, but hopefully you will get good criticism and be in the position to execute the plan if need be.

Now we will talk about plans that flop not because they are bad plans, but because the circumstances change which are out of your control. An example of this is when you have a contingency plan to have babies, but your group is active and all of the sudden you have three pregnant women. An abortion may not be illegal or legal by this time since the world or legal system in your country is wiped out by a super volcano. Whichever your view point on abortion is, you need to take a step back and look at the future. An abortion will have a high risk of destroying future conceptions especially if you don't have a medical expert that has performed one. Second, your survival will be based on how fast you can procreate. This is one reason you are here, so don't stop the babies from growing more than they already have. Okay, you were paying attention and you stored enough supplies for three babies all the way up to their adulthood. But you come to find out two of the women have twins. The chances of that are very slim, but you never know and biology in this case, not physics will usually prevail. Now you have five babies on the way but only have supplies for three additional mouths. Your contingency plan is a flop. Instead of trying the make the plan work, create a new one, and start making more food in the hydroponic room. Improvise and set up a laundry station in one of the bathrooms to handle the extra cloth diapers. Your plan should have included two years of baby nutrition supplies for the three possible babies, so you might be able to get by distributing the supplies among the five babies, having the mothers breast feed as much as possible and get the babies on solid food as quickly as possible.

The group of adults should plan on how many babies they wish to have while in the safe haven and when. This will hopefully keep all the couples from procreating at the same time. The same goes with possible teenage coupling where the couple falls in love and wed; or worst case scenario they procreate out of wedlock and now you have a single parent. The adults need to keep an eye out on situations like this and the teenagers need to be taught the dire consequences of such sexual experimentation and behavior.

I have been talking a lot about issues, but there are things like not having enough information in your library that you need and you end up recreating the wheel out of mental memory, assumptions, or word of mouth. With today's hardware and software technology you should not have a weak library or network of information. You should be able without condensing files fit 500 DVD non-blueray quality movies into one terabyte space of data. You can even create a server system and have ten terabytes for all your information needs and be able to supply the group with information for movies, shows, games, books; everything you can normally do on a computer, tablet, or cell phone with the exception of having internet access to a world which has been wiped out. With technology moving up in capabilities and going down in price, ten terabytes of hard drive space should cost no more than $500.

I'm not going to give you a set amount of data space you should have, but remember that whatever amount you come up with, double it for backup purposes. I recommend you have enough space to be able to give the impression that it would take years for someone to see all of the videos in the database. In addition; content of information should be approved by the group of adults and monitored so that the very young children do not get exposed to more mature subjects. Also, the use of visual aids like flat screens and projectors are nice, but make sure that they will last and you have the ability to interchange components.

The contingency plans will be stored in this database and everyone needs to know how to access the information. There might be information you don't want everyone to have access to, which can be taken care of by good network management and permissions. This information I am talking about is <u>graphic</u> step by step information on surgery, sexual in nature for teaching the children when they are of right age, giving birth, or news clips that depict mature content, and information similar to Last Will and Testimonies of the adults for the children and group. You could have all this on paper, and you should make storage space for critical paper data, but videos are best to convey the last words of a deceased and how to procedures.

The plans should all combine to make a well rounded plan. Now, don't confuse a reactive drill with a contingency plan. A reactive drill is something you might find in a contingency plan, but it is not a plan. The reason is that a drill is something that you automatically do in response to a situation and is not a fix for all problems. A contingency plan is a solution to a problem. A contingency plan is an adaptation to normal and abnormal operations. For example: if your solar panels are disabled due to fallout, Sun blockage, equipment damage, or physical disconnection, then the contingency plan needs to address alternative ways of getting energy into the batteries and provide a steady flow of energy for the critical systems, while you fix the solar panels or increase the energy production of an alternate source. Ideally a nuclear power source or a source similar to a dam would alleviate a need for a complex contingency plan to address an issue like this. Having said that, it would help greatly if your initial foundation is reliable, easy to fix, and redundant. The intent is to reduce the need for contingency plans.

The unknown is something that you can try to plan for, but not in a contingency plan. What I mean by this is you can have a contingency plan for losing a visual on your security area or over watch on your obstacles; but a contingency plan will not work for an unexpected freak of chance that a Being 747 crashes into your farm/food source during a food crisis situation. The sprinklers that were in your contingency plan to fight against a fire into your farm doesn't really do the job on jet fuel and wreckage. The survivors and dead bodies which your group should morally be inclined to help and feed even though it might worsen your chances of preserving any food source not already destroyed cannot be placed in a contingency plan. Why? Because even though I told you to come up with all possibilities, there is a point where you will not be able to cover everything. Your leader and the adaptation of the team will play into acting promptly to situations which over shadow any reactive drill, contingency plan or base plan for survival.

So what do you do to cover all the areas? Make your base plan broken up as I described earlier in the book, then gradually make your contingency plans. Tweak everything as you prep, and once your safe haven is completely self-sufficient, go heavy on drills, individual and group training. **The proficiency of the group in following orders, taking the initiative in the absence of orders and collectively adapting and working together is what will cover the unknowns.** Much of this proficiency has to do with each person in the group knowing their jobs, what is expected of them, and knowing what everyone else is supposed to do. If the person who is supposed to secure the safe haven entrance gets inside and faints due to lack of oxygen, whoever is nearest should automatically alert the group and take on the task of securing the door before more toxins come into the safe haven. This is not a drill, this is a reaction by an individual or group to take the initiative and ensure critical survival tasks are accomplished. If it were a drill, an alternate would have been designated and standing ready to take on the task, which was not in this scenario.

The well rounded plan must have everyone included to the point of being able to live without the luxuries of a world that is not filled with toxins in the air, rioting people, or many other problems like no internet connection which includes all that television/radio transmissions many people have grown up with. Fears need to be addressed early on so that they don't have to be addressed while in the enclosed area of your safe haven. Once your safe haven is self sustaining and you are in the executing phase of your prepping, you can practice living in close quarters.

Astronauts train many months if not years in enclosed environments on the ground before they go into space. Many people might love staying indoors, but most on occasion go outside to run an errand, for work, or whatever, but when you are forced to stay indoors and not even open a window, then the mind thinks and feels differently. If you plan well and take into account all of the possible everyday lifestyle events which will probably occur in your safe haven, you will not have social or psychological problems with people going crazy (this includes anxiety attacks) once they get put into a very stressful and unyielding situation.

This is another reason why it is important to have a large database, collection of games, books, and things to keep everyone active individually and as a group (not busy work). Activity could spell watching movies or shows, as long as there is a routine which includes good nutrition, sleep, and exercise. Otherwise watching videos for six hours a day is possible, not recommended, but possible without the person getting fat and lazy. Like I said earlier, you want to have routines that are closest to your everyday lifestyles in the past. School and training or learning a profession while in the safe haven is something that will keep a person active and their minds away from unedifying negative thinking. The plan must account for these desired routines. For many people, not having to go to work will be a shock or blessing now that they can spend every day with their family. But, it could go against you. There have been studies where high ranking enlisted military personnel who after they retire end up dying ten years later. The retirees who have a mirrored lifestyle close to the pace of their life long career end up living a lot longer than ten years. The shift from work to a slower pace will cause your body to react in negative or positive ways. It might be the best vacation ever for you, once you get inside your safe haven, and you don't have to worry about anything except your family. Many of the stress producing factors in your life (like a job or people you are forced to associate with) will be gone if a super volcano erupts, but in its place will be the survival factors which will bring in their own type of stress.

There is good and bad stress, and your plan should maximize the good stress and allowing for events which address stress management. Your plan should include effectively performing everyday activity like washing dishes with stress relieving measures. It could be using a dish washer, or intentionally washing by hand giving the children a good work ethic. In essence, focus on things which produce responsibility building activities and nurture love, respect, and leadership. Humor is one of the best ways to manage stress and keep people motivated. Your plan must have humor and laughter in the midst of the crisis; without it, you will live in your safe haven stressed and defeated, looking at yourself and saying poor me. You can have comedy shows, but be careful because not all humor is good for children and even adults. I recommend you stay away from humor which points to negative racial, gender, or sexual aspects of people. Bill Cosby as a stand up is good at this. One major issue with comedies is vulgar language which should be strongly censored. Children may grow up looking at "F" "B" "A" and "S" words which really don't need to be seen or taught as comical. An example of what not to have as humor relief are shows like American Dad, Simpsons, Archer, or South Park, this is only the tip of the ice berg. There is edifying humor, just be creative and you will find it. Flintstones, even some Friends episodes (not most), My Three Sons, Father Knows Best, Leave it to Beaver, I Dream of Genie, Abbot and Costello, Cosby Show, Charlie Brown, Fat Albert, Fred Astaire and Ginger Rogers, and much more. These are old shows, and yes they are, but the quality of humor you want should not be little others, focus on sex, or insinuate inferiority of others, even if it is not anyone in the group or an entity.

To get back on track with a well rounded plan, your

database should address documentary information. Your children need to see the wild kingdom side of reality, even though many of the beautiful animals they see might not exist once they get out of the safe haven. All the documentaries of science and history should be a part of your well rounded plan. Video clips of WWI and WWII are good teaching aids for the teacher, but in some cases they should be shown when everyone is present. That could be a group project for the children to learn about history, why things happened, and what could have been done better to keep things from happening again.

The movie, "Reign of Fire" depicts the leaders of the castle playing out the Star Wars Empire Strikes Back scene of Luke and Vader where Dark Vader reveals he is Luke's father to the children. They only had the knowledge of the actors to tell the story instead of a video. You can role play things to, like in real plays and get people involved with the stories. Or simply show movies on the screen. Whatever, the case, your plan needs to address the needs of the group for information. If you don't give people something to think about all they will be thinking about is what is going on outside, how much someone else annoys them, or fall into a routine that is emotionally depressing.

Your plan must address all the needs of the group mentally, physically, socially, and spiritually. It must also address desires of the group like eating ice cream, cake, privacy for diary writing, able to pursue a talent, or simple things like watching favorite shows or reading favorite books. Ice cream and cake might be an issue since ingredients like eggs might not be available, so you have to get a good substitute like powdered eggs. The larger the group, the more diversity you will have which is what you want to some degree. If diversity leads to unresolved conflict then the group dynamics is not the right mix, and you should find different people.

There are also needs for survival which can cause problems but should be addressed early on. As I mentioned earlier, what do you do if a person in a couple become unable to procreate. Well for the good of the couple the female needs to be impregnated or the male needs to impregnate a female in the group. What you don't want is for physical mating to occur between the non-couples. Your female medical expert needs to know how to artificially impregnate the females. The couples also need to understand that the mother and father is the couple, not the odd couple in the group. It must be stressed in the plan for the survival of the families that this is known to everyone, that the cells came from Uncle Tom for instance but Sam is the father. Being the biological father or mother does not give the right for that person to claim the child as theirs. In court it might to some degree, but the agreement is for the survival of the group, and the families within the group. You can have disputes or resentment stir up because of the emotional or biological bond, but remember that the gift of a life for someone else is more important. There can be all sorts of what ifs on this subject, but my point is you need to address all of these situations that might destroy the group or an individual, from the start of making plans.

I talked about the group in California that had a no weapons policy. They didn't have a well rounded plan and still don't. Why? Because; an axe is a tool which they have, or a hammer for that matter, and if things hit the fan, and it gets ugly in the camp because squatters have taken over, those tools are also weapons. What happens to the policy when one of the members kill a squatter or squatters? A policy is just what it implies, a rule which is to some degree a guideline or in the extreme a "law". People don't go to jail for breaking polices, they might get fired for not following a policy, and go to jail for breaking the law which might be a policy or policies. So, maybe the group should reword policy to law. Otherwise they will be playing with words back and forth when the important things aren't answered properly. The feeling in the leadership was God would protect them. So, what does that mean really, do you do nothing at all? Do you create a bunker or safe room? Do you get non-lethal weapons or methods of dealing with people who may not have weapons, but they do have a desire to take what you have. Their plan should have addressed procedures or policies for taking in or deterring drug addicts, marauders, or things like patients from a nearby asylum who might try to kill you because they think you are a demon or monster. If they are willing to accept people in the fold with open arms since they don't have weapons to enforce law and order, then have they planned for mass casualties or hordes of people from an exodus of the cities? Probably not, and there lies my point once again that a great plan has to be and is well rounded when it doesn't limit things that keep order and work for what is best for the group and individual.

Now I must talk about Doomsday Castle. I applaud the family and the setup of the castle and all; however, there are things which were not planned in addressing critical situations. The castle is very tall and all constructions have their pros and cons. It is a big target, easy to see and find; granted their saving grace is that it's surrounded by forest and rolling hills. A tank would have a very difficult time getting within firing range of the castle, if not impossible, had it not been for a road giving access to the area. A tank round will take out the castle, as would many things, and not many things cannot withstand heavy military munitions. My point is that they focused on having a draw bridge, catapult, firing ports in windows, fire resistant measures and all, which was cool to watch; but since it is in the middle of the woods and on a hill. The <u>probable</u> human enemy they might have will be dismounted personnel. These people will likely have weapons as mentioned on the show, but do they really know what a dismounted platoon size element can do? The father in the family said he was an Infantry instructor for four years in the US Army. It shows and he is an expert in many areas, but so are many people like myself; who spent eight years in the infantry as airborne, mechanized, light infantry, Ranger, and JRTC OPFOR; and the other years in counterintelligence. There are many people who never served in the military but they are survival bred hunters and can shoot weapons just as good or better than most military personnel. So, what do you do when a platoon sits on the surrounding hills and waits for someone to come outside to hunt or expose themselves within the castle and they start getting picked off by snipers? In essence the enemy can siege the castle from a distance until the occupants are reduced to such numbers that defending it will be impossible. I'm sure the father would react appropriately and snipe back, use indirect fire on them, wait until night and see if they can find and kill the snipers. Several good options are there for the family. They have access to explosives which means they can make very effective and deadly direct and indirect weapons. The thing is this scenario is not in the plan even though they had people firing paint balls into the castle and people died. The father will know exactly what to do, but what if he is killed on the first sniping attack. Someone else might know what to do, but then it is a matter of instructing the rest of the people on the strategy or responses. This is not the time to learn how to repel probable hostile actions. And if they are not prepared for psychological warfare in the form of slow and purposed killing, then they are not ready with a plan or drill that is beyond a small or large conventional attack as they exercised on the show.

The military trains on battle drills and there are a lot of them, but the castle is a catered situation that all of the family members need to know. This is part of the plan which must address the most likely to the unlikely scenarios of a company or larger size element coming into their vicinity and wanting to cause them harm. The objective of the plan must get the group to a point where threats are deterred, eliminated, or made non-threatening. The plus that the family has is that if it looks hard to conquer, the enemy will not bother trying to take it and move on to easier targets. So the family did really good by making the castle large and impressive to look at. It might also even help them gain allies. If I were a group moving around, I would make peace with the people inside, which is taking a chance, but at the very least I would make verbal contact to let them know who is in the area so they don't shoot me by accident, and at the same time I can get a feeling of who is inside. Which as I address in this book, you should know who is around your safe haven at least 30-

100 miles around you. In addition, your plan should cover possibilities that due to the crisis people who were your friends or allies in the past are no longer your friends or allies now.

Having said all of this about a well rounded plan, the rule of thumb is your planning should look outside of the box. You should not be leaving out of your safe haven area or bunker during the entire time of the crisis. But since you think out of the box, you plan for leaving the area, so that you won't be sorry later in case there was an emergency where your group would die if you didn't get outside help. You create a furnace plan in case a member of the group dies, this includes animals. A furnace might attract attention, but it is something necessary if you don't want a dead corpse to bring you more problems than attracting attention. In a self-contained safe haven, a furnace is a necessity, not an option. If you are not in a self-contained environment and have access to the outside, burial is an option, but you better dig the grave deep and far away from your water source. The deeper, the better; recommend six feet or more. If you dig the grave too shallow, animals will dig it out or expose it. This will cause problems with attracting attention to humans and other animals, in particular vultures and insects which can contaminate things in your property and attract humans looking in the sky.

As you can tell, I have jumped around a lot from subject to subject. Like I said before, many things interlock with each other and your plan is how you see the big picture, the details, the gaps, and most importantly where you are headed. I will talk later about meetings and deciding things as a leader and group. Your great plan should guide you to the kinds of battle drills you will train on and be familiar with, contingency plans you will likely use, and how the prepping and executing comes together.

Chapter Eight

* | * | *

PLAN FOR NEXT YEAR, NOT FOR TOMORROW

We talked about covering all the ground you can cover to include the unknown, but you must keep your focus on living, not just surviving. Science has shown us a world full of abundance and we think that it will be there no matter what happens. Even though science also screams that full of abundance has a limit. The Earth's resources are being depleted faster than we care to admit or faster than we are able or willing to replenish. The supplies which will help get you through a few years is not your ticket to survival. I told you about the movie where the young lady goes outside of a fallout shelter at the end and finds a world completely destroyed by a nuclear holocaust. The survival success of you and your group is based on what will happen to the future generations of the children. You might have two or five years of supplies, but there must be a secondary Ace in the hole. If I had time and resources I would create a network of preppers and a complex of safe havens in key locations. The safe havens should be within two miles or closer apart. The equipment which you get to monitor the air and radioactivity outside of your safe haven must be reliable and able to be portable when your recon team goes out to make contact with other preppers. If there are no other preppers to make contact with, your recon team should have an objective to find a good water source and area that can produce food. It might be a cache which you strategically placed so that you can create a farm or drill for resources. The intent is to expand from your safe haven into the world and repopulate and start civilization again. As for an economic crisis it is to rebuild and organize a strong economy again. This might be harder to do because the economic collapse may bring consequences that will fight you for a very long time. Another country might try to invade your country or region. Records of land might get lost or be deemed non-binding if you whip out a deed for your property. The bartering of the land might be crazy and many people may try to constantly steal from you or attack you once you leave your safe haven area. If you kill someone protecting your safe haven, the established authorities might deem you criminals so now you will be fighting law enforcement before you can straighten things out with them, or get higher authorities to clear up the situation. In worst case scenarios you might go on trial for murder, probably manslaughter, which might happen with the group in central Florida if they persist on governing themselves when the country has a system. The system will be based on who is the strongest entity in the land.

Videos do wonders nowadays when it comes to criminal cases. The security and surveillance of your safe haven maybe your witness of what happens around you. I highly recommend you have video cameras in all the vehicles you have in your safe haven. You can buy video cameras that fit in your helmets so when you are in a combat situation the command post (safe haven) can see what it going on outside and inside through the eyes of the adults maneuvering around.

The goals of your plan will cause you to plan for the distant future. In all events, you must plan on sending out a recon team to make sure things are safe when you have decided to leave your safe haven area. How you conduct that recon is different for all preppers due to the situation because there will be different requirements for each group. In an economic crisis, the requirement to monitor air quality and water may not be as intensive as it would be after a nuclear war or super volcano. If your safe haven is in the mountains as opposed to the swamps your mode of travel, equipment, and consumables will be different.

There are some common things which all recon teams need to have. They all need communication in the form of radio, signals, and video if possible. You can have texting, but it is not the preferred method. They all need weapons, which can vary in types, but they need to have three categories of weapons, which are silent weapons, heavy suppression or assault weapons, distant engaging weapons, and weapons of mass confusion (referring to smoke, large area stink bombs, CS, or fire bombs). They all need to have a mode of travel which gives them mobility and capacity to survive an ambush. Lastly, they all need to have the ability to navigate without getting totally lost and carry enough consumables to get to their objective locations two times over.

Communication is quite easy with today's electronic devices in the form of hand held radios with ear pieces/Bluetooth. Secure and unsecure capabilities are available, plus you have digital video cameras which can act for recording and real time reporting to the safe haven or even within the recon group. There are robots premade and homemade that can search areas to make sure the area is clear of radioactivity, toxins, and booby traps on a road or a door. Apart from radio, there are hand and arm signals within the recon team, which include flags, colored smoke, and flares. People in movies say stylish code names or words while on the radio, but that is not shown correctly most of the time and can be become bad practice if done improperly or for no reason. Secure communication does not require code names, brevity words, or confusing extra words. The less you talk, the more battery life you give your radio system. Speaking real names and simple conversation should be used if your comms is secure. If you don't have secure comms, then you can use codenames, brevity codes, and the like, but chances are that even with non-secure comms your conversations will not be compromised. Unless you use a cheap radio which only has a few channels and the enemy have the same radio or access to those channels; which is unlikely unless you both went to the same electronic store and they know who to change the frequencies to match your radio band. If you don't have comms with your recon team, you might end up allowing the enemy enter your safe haven property because you think they are your team. You might also don't have real time information on what is going on around you, which can get someone killed. An example would be your team is moving around outside and they get lost and enter your booby traps or you see someone in the area and you launch countermeasures like a fire bomb. If you have no comms with your team, they will get attacked by your safe haven and probably injured badly before they can signal you that they are friendly.

The weapons which must be carried are based on the mission objective of the team. I stated recon team, because that is what the objective should be. It can be a team that conducts an ambush, raid, or assault; but then it would not be called a recon team. A recon is to gather information without being detected. What I mean to say with recon, is you need to have a team that starts out on a recon mission and hopefully ends up that way; however, they need to be flexible. The silent weapons are so you can act without giving yourself away. Throwing weapons, a tomahawk, knife, and the like are good. Silencers are illegal in some states, but I guess if a super volcano occurs, you can go into your deep storage and break out the homemade silencers, or ones you acquired from military or police channels. If you have a silencer or any other illegal weapon, and you are expecting an economic collapse, I highly recommend you get rid of them or don't use them. If they are legal in your state, then just use them wisely and safely. There will be authority and laws in the country and having illegal things like that can cause you more problems than it's worth.

Assault or suppression weapons are a must in case you are ambushed or run into a large hostile force. The best for medium range combat are rapid fire shotguns. AR-15s, AK-47, or M4s are best for beyond a hundred meters to three-hundred meters. A M-

249 (SAW, or Squad Automatic Weapon) type of weapon are best for all-around ranges and targets. Pistols are best for targets within fifty meters. Not that they have the penetration power like the rifles, but they fire rapidly, can be manipulated by the firer easier, and can be employed much faster than rifles. They are also best for close quarters where there are a lot of corners and rooms. If you really want that penetration power, you can get large caliber pistols with armor piercing rounds. All of these weapons sound great, but there is one consideration you must always take into account when deciding what assault weapons you will take. Ammunition is a large load on any person. 180 rounds of 5.56mm M-16 rounds is roughly seven pounds; which is six 30 round magazines. If you get into a firefight six magazines will not last for a long time. The more rounds you have the better, but that means people will carry more weight. 7.62mm rounds are heavier and well you sort of get the point that if you have a weapon that consumes a lot of rounds you run into weight hauling problems. If you have a vehicle that would help, but then you are not going to be very stealthily or effective if you only have one vehicle. If you have a donkey or horse, that might work better, but there are still areas even horses cannot go like up a mountain or in this example cliff. Horses and vehicles are also easier to track if you are being followed. My point is, be aware of the loads and things which hamper or assist the team.

Sniper rifles of course are best as distance weapons. A 308 is generally more accurate at certain distances and easier to handle for the average novice. You should see what is available and what best suits you by firing sniper weapons. I recommend a sniper rifle that can reload quickly, and is light to carry, but has high penetration. Practice and more practice will help you hitting stationary and moving targets, but your snipers need to know about cover and concealment. Not that you have to be wearing a Gillie suit in the heat or cold of the day, but your sniper needs to know the trades of the sniper in order to be able to shoot at other snipers, and also be effective as a sniper. There is a lot involved with sniping, which is why there is professional training in the military and law enforcement agencies, to include on the job experience which make snipers the real professionals they are. But as a novice, knowing your weapon and practicing by firing as much as possible is a good start.

Now down to what I call weapons of mass confusion, which is normally labeled riot control weapons. The fire bomb is not riot control, but is a measure to kill and control at the same time. Having said that, your weapons should include concentrated skunk bombs or sprays; pepper bombs or sprays; or sound disruptors. The US Army likes to use simulators during training exercises to simulate artillery rounds or large explosions hitting around your location. They are very loud, and people are injured if you are standing too close to one. If the simulator is on top of you when it goes off, it can cause you to lose a finger or worse; but don't look at it as an explosive to do physical damage. What you should focus on is the concussion effects it represents in your favor. I will talk more about fireworks as a means in using little resources as a substitute to expensive explosives and weapons in a later chapter. Right now, what you should consider is finding and using hand grenade type weapons, like smoke canisters to keep the enemy from easily targeting you. Something you can carry and deploy quickly, as opposed to having to pull out a large canister, light it, and then throw it. There should be a device like normal hand grenades that you pull two safeties and then throw. A grenade that is loud enough to deafen the ears of people within a twenty meter radius is something you want to carry. It might attract people in the far distance but it will help you get away or disorientate the enemy for you to get the advantage. Most people wear body armor, but most people do not wear hearing protection. At a firing range they do, but in combat, not so much. If they do have hearing protection, then it might not be that effective; however, it will cause the person to duck or look around and look for what made such a loud bang. It is a thought, get creative and you can come up with good throwing weapons or weapons you can sling at the enemy that will cause confusion or panic in their ranks.

I placed this information in this chapter, because you are trying to prepare for the future, and these things I have covered are mainly for when things are over and you must go out into the world. Your children need to be taught these skills and if you have the resources, I would make an underground indoor range to go with your bunker so that you can train anytime without having to go outside. If for some reason you are in the safe haven for a ten year period and come out to the surface, you will have to conduct weapon training outside and you don't know what might hear you. An indoor range can also be used as storage or playground until you need to use it for shooting ammo down range. Your young children who weren't able to train on weapons will have to do it at that time, unless you have an indoor range. There are weapon simulators to include PS3 and Xbox, but there is no real substitute for a real weapon with real ammo; without a major expenditure for military or law enforcement weapon simulators.

Experts say that you should have three forms of food production and all the other areas for survival. Keeping this in mind as I have talked about backups, you need to make sure you are not looking at just training your children to know all you know in survival, you should be focused on that plus, how to become a man or woman. Teach them the things which gives them character, gives them parenting skills, and goals for themselves after the crisis. If I am a leatherworker or doctor for instance, I may teach my daughter my trade or skills, but what will benefit her and the group the most is if she learns about life, becoming an adult, and having a desire to live and contribute to the group and family. What you don't want is to force or even sneak in your trade into her life, because that is what you want for her. You might think - what is wrong with that? Well, if the youngster knows that he or she doesn't have to follow the footsteps of the parents or someone else, but decides to because they want to and like it, then there is nothing wrong. However, if the child has not been exposed to other options, and is not mature enough to know what he or she wants to do, then it will have a chance of depressing or giving that child/teenager regrets later on. Building character is the name of the game when it comes to prepping for way down the road.

Lastly, if you plan on coming out of your safe haven and you conduct your recon, let's say 100 miles out in all directions; and you find nothing worthwhile or other people. I strongly recommend you use what you have and make your stake where you are. Unless you can see that there is a settlement of people that you have contacted by radio and preferably have video confirmation on who they are, then go to them even if it is a thousand miles away; if you have the resources to make it there. Otherwise, the grass is not greener on the other side. If a super volcano destroyed the area around you, chances are it destroyed the areas around the other areas around you. Stabilize what you have in the group, make the group stronger by numbers, start a town or city if you have too. In time when you have the resources of people and more machinery, then you can afford to send out parties to look for other civilizations. Or maybe they will come to you. My point is, don't stretch your small group out because the crisis has abated and you think there is greener grass beyond the ocean, sort of speaking. If you know you can't survive as a group in the location or environment you have stepped into, then moving the entire group until you find resources or other people might be your best option.

Chapter Nine

* | * | *

WHAT TO DO WITH LIMITED RESOURCES

There are many problems that arise when you want to do something, but you don't have the resources. The resources include money, time, information, moral support, strong relationships, hardware, software, food, water, electricity, good air, and connections. You have to get creative and do research. The rooms in your safe haven which I talked about earlier do not all have to be as described; however, you don't want to store your ammunition with your weapons unless there are safe guards to prevent one person from getting to both and going postal on the entire group. Don't think that it is impossible. It has happened in the military where a Soldier gets a letter from a loved one, and became so depressed that he/she wanted to commit suicide. In the process of going to the arms room, he/she loaded ammunition which was being temporarily stored there and started shooting whoever was around before attempting to shoot themselves. The military has strict policies about ammunition and weapons storage. They also have strong procedures for storage of flammables which are stored outside in separate lockers or sheds. My point is that there are shortcuts you cannot compromise because it is hazardous and can get you killed before the crisis even happens. This will be challenging if your goal is to an underground self-contained safe haven, but it can be done; however, this goal is one you cannot do without a lot of resources. In particular, the atmosphere control and disposal of hazardous materials.

I must say that if you are in an indoor environment, it's best that the responsible adults and teenagers carry weapons with ammunition, not chambered and on safe, so that the people are constantly ready and get into a good habit. However, do not get in the habit of carrying a sidearm because you are the boss, security officer, or other positions because it is part of a status. There should be a purpose tied into the threat; otherwise you will have people carrying weapons everywhere for bad reasons, instead of being ready when the crisis occurs or is obviously imminent. This is so you don't have someone get into an argument and use a weapon which can cause legal problems from the local law enforcement agencies. There is merit in training as you fight, so carrying weapons might be something your group needs to practice; however, be flexible and make sure weapon carrying it not made into a status symbol or right for everyone to do simply because they know how to use a weapon.

There was a scene in Doomsday Castle where one of the brothers carried a sidearm with him, in the back of his pants on the belt line. The other brother saw it and was like what are you doing with that, as if the brother was putting them in danger. I agree that the brother should have taken the sidearm with them for protection and such; but he should have taken a holster. The way he was carrying the sidearm was wrong and unsafe. They found a wild pig in their trap and it was jumping out of the pit by the time they responded to the bell alerting them of the trap being triggered. The brother had to shoot the pig several times before it died. If the brother had not brought the sidearm, I'm sure the frantic 250 pound pig would have seriously hurt one or both of the men. Always store things properly, to include things on your person or near you.

Your resources will be stretched and space will be your enemy with a lot or little resources. If you have limited resources, you might be able to make your living room the dining room, classroom and recreation room all in one room, but it will be cramped and it will be hard to do two activities or more at the same time. You can try to make your sleeping room to be occupied by the parents and children, but that is not practical and in the end will lead to more problems as the children grow older and the lack of privacy wears out everyone's need to privacy during the day and night. The little resources you have cannot be overlooked when it comes to getting the space you need. In most cases you will have to work on getting more resources when it comes to space and structures.

I mentioned that you can start slowly and learn what you need to learn and save money. Or you can be trying to get out of debt, which means you probably don't have the money to spend on things which you can't get without spending a lot of money.

But there are things you can do with the income or resources you do get. Canning food is a big one a lot of preppers start doing and most can do because it is done in conjunction with your daily cooking and eating. You cook for an extra person or more and can what is extra. You keep all the glass, plastic, metal, and paper containers and use certain ones for canning food, others for storing other items besides food, and others you use to create other equipment, tools, or to make money, like making crafts for sale. If you are a hoarder you might have a good start in this effort or maybe not. If you hoard lamps, trinkets, figurines, and things which really don't have a value for survival, then it all depends on what you can sell to make money or what you can use for things you need in your prepping effort. Some people have things like 8 track tapes. If you can sell them great, if not, then <u>unless you know</u> how to melt that stuff down and create something you need, then it is trash. People horde newspapers or books. Once again, see what you can sell, and the rest will probably be worthless and can be a hindrance in the future if you don't get rid of items with little value for your cause. Items or materials that last or is used in construction you generally want to keep. There are things that are expensive or inexpensive which you need to keep or get instead of thinking they are trash.

Here is a short list of what you should horde or try to get through your daily life. Keep in mind that if you are renting or your current residence is not going to be your safe haven; then you must focus on learning, training, and hording should a low priority until you get an established long term property. The intent of this list is to give you an idea of the items you should think about when it comes to things to keep or get – dependent on the crisis you're trying to survive. Small mirrors, compasses, Nu-wave Oven, hydroponics equipment, livestock like donkeys or horses (if you have the space for it like a farm or ranch), bikes, motorcycles (preferably cross country), wind mills and parts, solar cells, water purifiers, air purifiers, farming equipment, internal AC systems/parts, weapons (modern and past) like a Tomahawk – bow and arrow – catapult - traps, hand tools as backups for electrical tools, radios (old and new), computers (old and new), manuals of all kinds, sewing machine (electrical and manual), plug in vitals machines, minimize use of batteries and storage of batteries (your goal with batteries is to use rechargeable batteries whenever possible), and all kinds of metals which you can use to sell as scrap or incorporate them in your home.

Many old items can be refurbished or used to create a better item which can substitute for modern items like many radios and communication systems. An old style PRC-64 radio is not pretty or better than most modern handheld comms between the group; but it does the job and is likely not going to be eased dropped on. A hording of walk-e-talkies may not be pretty and is not the best comms for minimizing sound and security considerations; but it is something that will help the group greatly if everyone can talk while performing tasks in construction, training, or tasking people around instead of hunting people down to get things done.

Items like explosives, chemicals, special fabrics (like heat retardant or acid retardant fabrics), can be horded with minor expense. There are legal items that you can gather to create a warhead for you rockets, fire balls, and obstacles. This includes stable chemicals or grains that once you mix as a paste or in liquid form act <u>like</u> C4. Your goal is not to create improvised IEDs that will take out a truck or tank coming down your driveway. If you do this, the government will have justified grounds to treat you as a terrorist even though you are intending to be one in the middle of martial law or civil unrest. This is one reason Waco Texas became an issue. Not that I am siding with any side, but the alleged activity by the Branch Davidians is what drew the long arm of the law in the form of ATF and well things got escalated to lethal proportions by both sides. I will talk about deadly force later, but just keep in mind that acts of very strong violence will be met with deadly force by law enforcement or military if there is still law and order, whether it is our great current Constitutional system or another type of state or national government.

Now, if you are or can find someone who works with fireworks, you can get valuable information or resources in making the exploding part of the missile that you can use to fire at people and vehicles. The loud boom everyone hears during fireworks is what I'm using as an example. You can get information on building rockets out of wood, plastic, and metal. It is a big hobby and big business; but my point is you can develop/create a rocket launcher or a cluster of missile launchers. Understand that rockets and missiles are two different animals, not by much but enough that they are not the same thing. The warheads of these rockets or missiles don't have to be high tech, military grade, tank buster type of creations. The concussion of a firework rocket exploding next to a person will deafen that person, temporarily or permanently. If you can pack enough concussive force, no matter how much hearing protection the enemy has, it will not keep them from being injured or frightened. The US Army used to have a heavy anti-tank weapon called the "M-47 Dragon". It was created in 1975, and was known for its ability to destroy any known tank on the battlefield. It is interesting to note that the impact of the round on a tank or ground killed dismounted personnel or ruptured ear drums within a fifteen meter radius of where it hit. I used to be a Dragon instructor in the Army and have walked around the impact areas of the rounds. I found dead birds twenty meters away, and they just didn't die, they practically popped like the birds in the movie "Shrek". The concussion was so great that the lungs are collapsed and just simply devastated the inside of a body's internals, not to mention the ear drums. So, with limited resources, you can be creative and make weapons that don't necessarily have the devastation of military munitions, but they can come close and close is all you need to get people to think twice about continuing to fight you.

Items like bleach should be stockpiled; however, you need to research the best way to store chemicals or items that degrade with time, because bleach in liquid form will not last as long as bleach in solid form, depending on the packaging and storage temperatures, etc. Some liquid forms are more hazardous than solid forms when it comes to storage issues. Once you figure out the forms and storage needs, you can horde those particular items instead of wasting time stocking up many of gallons of liquids that in time degrade and become worthless. Almost everything has an expiration date. Meals Ready-to-Eat (MRE)s are used by the military and survivalist, which are great and do last a long time; however, they do have an expiration date and can be contaminated or spoiled. There are particular MREs that are likely to spoil, like tuna and chicken meals. So, when you are hording or looking for things, make sure you don't get things that are already expired, about to expire, or not fixable by you or someone you know.

The rule of thumb for working with limited resources is, be creative and start with what you have. But in the end you will need to get more resources by an increase in income or outside help from family, friends, organizations, and other preppers.

One of the best ways to increase your resources is to get with other preppers. They have similar goals as you and even if there is a falling out, chances are both you and they will have learned something like a new skill, better methods in prepping, more connections to other preppers, and many more. You can look at the negative aspects of why the falling out happened, but learn from it even if you think it wasn't your fault, pretend it was, and plan on it not happening again. Then focus on moving forward. Your limited resources are to some degree how much you limit yourself and what you think you can or cannot do.

If you have a mindset to use disposable diapers as one of your primary supply item, then you should reconsider. Disposable diapers are convenient and may be practical in an emergency, but they should not be the primary form of clothing babies. I am using this example so you can see that the mindset of what and how you use resources will determine if you are using limited resources effectively. Keep in mind that diapers are not just for babies; the elderly, sick, or injured adults and teens may need to use a diaper, so cloth diapers are not practical for these situations. The mindset should be to think about how your resources will be used, like in the diaper example; where I recommend you use both with cloth diapers as primary and you have disposable diapers in storage.

A dishwasher uses electricity and some people might find them not worth having because they can breakdown or use up too much energy which is coming out of your power sources. If you hand wash, it is possible it can conserve energy, use more water, or gives a person an activity. There are pros and cons for everything. With limited resources, I would try to get a hand me down dishwasher or take the one in your normal residence. Hand wash until the crisis occurs or until you can buy another dishwasher. A dishwasher will converse more water than most people who hand wash. You free up a person for at least an hour a day if you combine all the washing time. I would, looking at it from a resource perspective do both. If the dishwasher breaks you can use it for other purposes or just fix it. You want to have both mechanical/manual and electrical resources at all times.

The use of washing machines and dryers is as I have said above something you want to have instead of being sorry you didn't. You need to do both: use the machines, and hand wash clothes and air dry if possible. One consideration you want to always look at is if you are in a self-contained environment (underground and air/water sealed) the furnace, dryer, and all other devices to include open flame and chemicals will cause problems with your atmosphere. However, if you have the filtration systems, purification systems, hydroponics, plants, and water systems that create oxygen or food in the form of fish and plants, it will allow you to get away with using several of these devices without any issues. Remember that I am talking about the best case survival situation where you have to be in the ground and in your own artificial environment. Now if you are trying to survive the notion that there will be no electricity or parts for a washer, dryer, dishwasher, computers, and many more, then you are ignoring the ability to create energy from many sources. One of the main problems people run into is we have become too dependent on the local power company and power sources at your job.

I talked about the Nu-Wave oven, and this is an item which is the perfect substitute for the oven, oven toaster, and microwave oven. A microwave oven destroys all the nutrition in your food, and I highly recommend you never use one, and if you do use one, use it for 6-9 seconds to destroy bacteria, or as a substitute for baking clay or glass crafts. Otherwise the microwave oven will be an undoing for you in the long run. A Nu-Wave oven does the same things as a conventional over, but even better, faster, and a lot safer without open flames or large electric grills. The cost is less than the average 3.1 cubic foot microwave, and of course a lot cheaper than a stove oven. As for open flames, avoid open flames, especially underground; use an electric stove if that is your only choice, if for some reason you ignore the Nu-wave oven recommendation. Note that I never mention using natural gas. It is a great and cheap resource, but in an earthquake and other doomsday scenarios, you want to stay away from a possible time bomb in your home. The other major problem you have with natural gas, is that you will most likely not be proficient or knowledgeable in fixing gas lines or issues which are more dangerous than fixing a windmill, solar panel, generator, or watermill; and you shouldn't be expecting the gas company to come out to you and fix your problem in the middle of a doomsday crisis.

There are also many items that can be made manually or hand crafted. Most of these items however, do need resources in the form of space, materials, and special equipment. This actual creation of things like furniture, weapons, tools, and a wide variety of items like gliders or bikes can be created without you having to store the already built large item in a large space. You may not even need the item later, which is why you want to be able to fabricate items at will. The bottom line is if you can get a factory room for creating things that you need, you will save space in the long run and be more adaptive to the situation. The workshop is what will be expensive is all that it boils down to it. It will take time and effort to make a fabrication workshop, but in the long run your practicing or learning how to make things with your hands or machines will give you a big boost in your overall survival.

I said that the ideal location of a self-contained safe haven is on the side of a mountain. Keeping that in mind, there are gliders and other more noisy crafts and vehicles you can use to perform recons, or get your group down from the mountain instead of rappelling or making a path down a slope that is covered by debris. Of course this takes extensive resources, but it might also save you resources by using a glider to perform your recon. The glider will need a strip to take off and land. You can assemble the glider outside, and store it inside, disassembled. The glider can carry two people with gear and has a retractable jet engine so you can take off and get you to places you normally can't go with a glider. You want a glider because it is quiet and a good vehicle for areal recons. It's up to you if you want to make a helicopter or mixture of several crafts. My point is you should consider all possible avenues inside of just thinking about using a 4x4 rover to get you down the mountain or to conduct your recon. Like I said, earlier, knowing/being a machinist is a very valuable skill that opens the world of a machine factory into your safe haven.

Keep in mind that the situation of which you want to survive will determine what you should focus on first in getting or eventually getting. Note that items do not necessarily have to cover all of your needs all the time. An example is using a solar oven, which is an effective way of replacing a conventional oven without using electrical power. It is cheap and easy to make if you are a do it yourself type of person. If not, you can purchase one already made a lot cheaper than a conventional oven. The problem lies on the fact that you are dependent on the Sun, which means you can't cook at night, and when there is something blocking the daylight; like very dark clouds or debris from your doomsday event, you are oven-less without electricity. However, a solar oven is something you want to have, to use before, during, or after the doomsday event. You want to minimize the amount of energy you use in your safe haven, even though you might have an extensive energy supply. Remember what I said about habits, and conserving energy is always a good habit. It is not just energy, it's everything. Waterless urinals might seem like a hygiene issue, but if used properly it's not, and can be very good in conserving water and helps with waste disposal issues. If you do your research, you will find that there are ways to recycle things in your safe haven; it might cost machinery and expertise depending on what you are trying to recycle, but it will help in the long run to reduce strain on your waste disposal system.

The best thing you can have is elevation when it comes to a waste disposal system that drains down the side of your hill or mountain. You don't want a disposal system which the enemy can use to enter your safe haven or destroy your system so you are smoked out, sort of speaking. This is another reason you want a thousand feet or more above sea level and the base of the mountain.

I need to address financing for your prepping. There preppers that invest in precious items, like silver, gold, or clean water. The investment is usually not to become rich enough to build your safe haven. It can be, but most people would call it as investing in stocks to make money like every other stock broker. What I'm talking about is investing maybe $100 to $200 every month out of your paycheck into silver. I use silver because it is cheaper than gold and it is a strong commodity at the moment. If the economic collapse of the dollar or any other type o bill occurs, chances are that precious metals like silver will still be useable. You can sell the silver or pass it on to your children if the catastrophic event doesn't occur during your time. If there is a disaster, then you will have an item that is tradable. So, am I saying that silver will be worth something during or after the doomsday event? Well, yes and no. Depending on the crisis, silver might be worthless, and you will have a lot of coins you can melt down and use as a metal for other things. It is a very good conductor. So does this make silver not worthless? Let's look at this. If silver is useable as a valuable type of currency for items or services you might need after the collapse of the economy, then you might have made a good investment. **HOWEVER,** I really would like to warn you, because if you have read or will read about enemies, then having a Fort Knox in your safe haven might attract those enemies. You have to go outside of your safe haven and trade your silver or other valuable in order to get what you need like medical or food (as an example). This leaves you vulnerable to preying people. If you have an army – it might work, but then your army will be vulnerable to other possible armies. If the doomsday event kills so many people and the only ones around are other preppers then chances are that silver in this case will be worthless. **So,** what I'm trying to say is use your good judgment and stock up or invest in things that might be useful in the long run. Investing is silver is a good idea and I do recommend it since you can transfer it to the next generation. Another country might need it and pay you for it later. I would not advertise that I have it unless you are in desperate need of a life saving or life sustaining resource which silver can get you. Other than that, the best investment is to prep with five or more years of supplies, and the ability to be self-sustaining for generations, with silver or another commodity as a backup.

Chapter Ten

* | * | *

ROUTINES AND SUPPLIES

I am speaking specifically about these two subjects because they are your weapons against long term disaster. Routines, not habits, which you conduct while prepping and during your execution phase will either help or increase with the stress of long term activity or inactivity. It can also cause you to get into a careless mind frame and degrade the things you learned or teach.

The adults should have a routine of having group meetings. These meetings should not by all means follow a City Hall type of minutes and nominations, etc, like you see on television and in the real council meetings. The Army has a more effective system which takes the wasteful politics and imagined order in a meeting away. It is simple, everyone has a part and everyone is heard. What do they say? Well they report their findings to the leader. The leaders, in the ideal setup are two, the captain of the ship and the number one, sort of speaking. If you analyze this you see that basically four people report to the leader and the (two) leaders listen and ask questions. The four people reporting do not ask questions while they or their peers are reporting, unless the floor is opened up for questions by the leader. The leader and his/her second in command are suppose to make sure each person has their time for reporting whatever it is they are suppose to report on, and whatever concerns they have.

The second in command along with the librarian is supposed to make sure there is a structure to the reporting. If "Susan" for instance has the task of being the logistics and education expert, she would be given a format as to what she should talk about or cover. She and the group with the leader's approval <u>early on</u> decide on what should be addressed by her. In this case, Susan is given the task of reporting how all the kids are doing with their school work, how they are behaving, who seems to be having problems, who is doing above and beyond, what new subjects are going to be taught, and so on. The structure is based on what needs and what should be talked about or reported to the leader and the group. The security expert would report on breaches in the perimeter by animals, gaps in security, safety issues, equipment status on working and non-working items, etc. The meetings the military conduct are usually very long with a lot of detail on many levels, and sometimes information is simply said over and over daily, but in order for your routine meetings to function properly and be beneficial, they need to inform everyone with quality information. The security expert doesn't want to report that sunset is at 2034 hours, with a crescent moon, illumination of 90% for no reason. He/she needs to speak in layperson terms and say it will get dark at 8:30 PM tonight but the moon will be out with 90% illumination. We shouldn't have a problem with our night vision devices. There could be a chart or board where all pertinent information is placed like challenge and password, illumination, sunrise, sunset, chance of rain, key exams, status on damaged equipment, who is responsible for chores and things like that which will minimize the need to say things in the meetings.

The presentation must be simple. Many military people refer to many meetings as death by Power Point. The meetings are very long not because there is complicated information to be presented, but that at times too much information is placed on power point presentations which are colorful, animated, complicated to understand, or just have too many slides with non-critical information. You have to decide how your meetings are going to be presented. Slide presentations are very effective if done properly, and can give you a visual snap shot of what is going on, and what needs to be done. Now, you have energy consumption and equipment considerations which should not be overlooked. If you have handouts or a board to present your information for an individual or everyone in the group, that is great since you don't need to use up energy for a computer, screen, or a projector (easier to use a large screen television for a monitor for six people). However, the handouts use up paper, ink, and space. A board is good except you might run out of space or ink. So, I recommend you use all of them. Your librarian should be the one to help people update reports. In addition, if there is no time for a person to make a report, they can at least follow the template outline and address the key points the leader and group have already established as important. You brief by exception and talk to the group; if there is no time to put information on your report in the slides, handout, or board. All this sounds like a business meeting, and yes it is and should be treated as such. You wouldn't be lazy if your boss was expecting a briefing by you, otherwise you might get fired, so all the people/adults have to be diligent in all they do to include meetings, speaking and listening to children, teaching, learning, and performing hard labor.

In the prepping phase, the meetings are great to keep everyone informed on what has been accomplished, who needs help in an area, what is projected, what goals are expected to be done the next day, and so on. You can have more than one meeting a day, but I highly recommend you avoid that. There is such a thing as a group huddle which is a condensed version of okay who has done what and who needs help. This gives the leader and everyone else a quick snap shot of progress.

This necessary routine is for the group of adults, but also for the children. If you have teenagers, you can incorporate them in the meetings by having them attending one and later contributing by reporting information. You don't want to do this all the time because there will be conversations about people which you don't want the children to hear or find out through the teenager. Children will talk, especially young teenagers. However, this is a tool to get your teenagers in the fold and get them to think like leaders. In addition, you can have an entire group meeting session during dinner or after dinner, where instead of reporting mission critical information, the reporting can be on stories of what happened during the day or what is set in the calendar for the kids or the group. Like a fishing trip to the river, or camping and sleeping out in the tents. Remember, that if the crisis has not occurred and you are not in your bunker or in a defensive posture, you need to do activities which the children and even the adults will see as a vacation or bonding experience. Don't make it so all your experiences are based on survival training. Have fun and enjoy life in case you don't survive.

Routines are good and bad. Assigning tasks to everyone in the group is a good way of getting into routines, but it can cause what I call bad routines. Whatever you do, never, never, make a routine part of a punishment. It will help if you look at a routine as a chore. If you for some reason have a daily chore like burning poop from a portable latrine; don't make people that got in trouble the ones designated to burn the poop with diesel fuel. Not that they will go postal on you, but the fact that a chore is not enjoyable should not be the basis of why that person is doing the chore. The same goes with a chore that maybe one child loves and the other hates. The child who loves the chore, might volunteer for the chore or bargain with someone else to switch chores. The second in command of the group is the one responsible to monitor the overall distribution of tasks and responsibilities of everyone. It is okay for a child to voluntarily take on a chore, but not all the time.

Some routines are only for specific ages or genders. The entire group needs to know what they are. The males and females should separately teach and perform routine health or personal hygiene measures to ensure people's needs are being met without embarrassment or complication. An example of this is teaching females about menstruation, and the things they need to do. Females help females, and likewise the males have to address male concerns on the male side. I'm not saying adults should segregate everything all the time; dad might need to get involved with helping little Cindy age three to go potty. Adults need to determine what is appropriate and take measures to maturely get the group on the same sheet of music; but do not go around stomping your foot down with the Law or else type of attitude. I said this because many people try to create laws within the group to control too many things at one time. Basic guidance given to the group and knowledge of proper conduct should be followed up by all adults setting the examples.

Routines can also be looked at as a lifestyle. People all around the world wakeup in the morning and get ready to go to work. Some eat a good breakfast, some eat a poor breakfast, and some don't eat. Most take a shower and freshen up. They go to work, do their expected time, and hopefully go home and eat dinner. Or many do errands like pickup the kids from daycare, or hit the grocery store, etc. They try to relax or work more at home, and end up going to sleep for the next day of work. People do this five to six days a week, and then they use the weekend to relax. This is the average working class, and the average student has the same cycle of life for many years, except with new experiences as the years go by with new friends and teachers. The routines you have in your safe haven, and during your prepping, will be seen by your subconscious as just that. You will not look at going to your safe haven to improve the site as a vacation. And there lies one problem you need to correct. I'm not saying you need to get everyone or even one person to see prepping or being in your safe haven as a vacation or exciting adventure. No, it is work in many respects, but it is also a routine of home where you and your family live together as a family while working and when not working. Remember that I said, you need to make your prepping and safe haven reflect how you would normally live your life in respects to work hours, play time, house work, school, and the like. Make routines of life exciting, challenging, and constructive. Note that your work time in your prepping is also called training. If you have training which is not constructive, then you have not trained or worked well. Change up your routines if you see that they are getting to tedious on people. Move times around, get people involved on how to do things better or faster. Give incentives for accomplishments or even giving all they have in a good attempt.

Supplies

I'm going to talk about supplies now, but keep in mind that some routines tie into supplies. Canning food for example can become a routine, and it is important to know how to can, label food properly, store them properly, and have a precise inventory. The routines in tracking supplies, how you store them, and how you use up supplies will keep you from losing valuable time, supplies, and maybe lives. There is a 1993 movie called "The Last Outlaw", where the leader of the outlaw gang required the individual members to track their ammunition. He would say inventory, which actually meant status report, and each member would say if they were hit by a bullet or injured. Once each member said their portion they would go around again and tell the leader by round type and number of each round. The leader's expectation of the person knowing how much ammunition the person had in their possession was no joke. They redistributed ammunition among the group and continued with their 'movie'.

I'm not going to tell you the synopsis of the movie, but it was I think a worthwhile movie to watch and appreciate the routines and procedures the group had. The military is very similar to the outlaw gang. Everyone in a squad is expected to know how much ammunition and other items they have and used. The team leaders are expected to know serial numbers of weapons, blood types, etc. An ammunition check is always performed when possible after or during a fire fight. I use the example of ammunition because it is easy to see how important it is to know if you are low on ammunition when you need more rounds from another person or in your storage; or you need to change weapons or conserve rounds. Ammunition is a supply and you need to treat ALL supplies in like fashion.

There are differences however, when it comes to food or consumables, medical, barrier material, construction materials, fuel, or water. The US Army has a system of calling (labeling) these types of supplies by the term Class I, Class II, and so on. Below is a chart explaining the classes:

Class	Description
Class I	Subsistence (food), gratuitous (free) health and comfort items.
Class II	Clothing, individual equipment, tent-age, organizational tool sets and kits, hand tools, unclassified maps, administrative and housekeeping supplies and equipment.
Class III	Petroleum, fuels, lubricants, hydraulic and insulating oils, preservatives, liquids and gases, bulk chemical products, coolants, deicer and antifreeze compounds, components, and additives of petroleum and chemical products, and coal.
Class IV	Construction materials, including installed equipment and all fortification and barrier materials.

Class V	Ammunition of all types, bombs, explosives, mines, fuzes, detonators, pyrotechnics, missiles, rockets, propellants, and associated items.
Class VI	Personal demand items (such as health and hygiene products, soaps and toothpaste, writing material, snack food, beverages, cigarettes, batteries, alcohol, and cameras—nonmilitary sales items).
Class VII	Major end items such as launchers, tanks, mobile machine shops, and vehicles.
Class VIII	Medical material (equipment and consumables) including repair parts peculiar to medical equipment. (Class VIIIa – Medical consumable supplies not including blood & blood products; Class VIIIb – Blood & blood components (whole blood, platelets, plasma, packed red cells, etc)
Class IX	Repair parts and components to include kits, assemblies, and sub-assemblies (repairable or non-repairable) required for maintenance support of all equipment.
Class X	Material to support nonmilitary programs such as agriculture and economic development (not included in Classes I through IX).
Miscellane ous	Water, salvage, and captured material.

The military uses these terms of classes to address reporting requirements. For instance, if you are speaking in a meeting or on the radio you might ask when is class I? Everyone will know you are referring to when is food going to be served or is arriving. Okay, not that you need to know these terms, let alone, use them, but my point is that food and some medical items in particular are things which if not tracked properly expire or become contaminated to the point of contaminating other supplies in your shelf or room. If you don't store the items and know where they are stored, you might miss a batch of food that expires before another batch does and you lose food because you were not in the routine of conducting good management skills <u>by all the members</u> of the group.

Your supplies should be stored evenly distributed among areas in the safe haven to include your storage areas. Of course you want to keep the dangerous hazardous materials away from people and other volatile supplies. You also need to make sure you don't put all your eggs in one basket. This means you will have to create eight or more specific storage rooms. I recommend two rooms for food, two rooms for hazardous materials, two rooms for ammunition, and two rooms for non-consumables. Within a room you can separate things with barrier material or fire walls, in essence making more rooms or sections in one room. There should be a plant or hydroponics room which is not a supply, but it is sort of a supply of atmosphere. If for some reason there is a fire or something similar which will damage or completely destroy your supplies, you will not loss all of the supplies. You should have a vehicle or machine shop room, this is not a supply storage room, but like I said earlier, you can store anywhere you like, so I recommend you store things that you might use close to where you need it; in addition, you should store non-consumables where ever you can.

I highly recommend outside storage areas, which are separated from your safe haven by a distance of fifteen meters or more from the superstructure of the perimeter walls of your main structure. You should store items which you know you will not use unless you have to go outside or after the damaging crisis is over. Items would be bicycles, tractors, outdoor equipment, extra non-consumables, baseball and bat, and many more; whatever will make life easier after the crisis. In connection to this outside storage, I would make a deep storage area, which is only accessible through the safe haven, hidden, and is locked and sealed. If the seal is broken, then you know someone went into the storage room without authorization by the leaders. The items in the deep storage are the illegal items you decide to create or get. These items are but not necessarily illegal in all states but include silencers, flame throwers, lethal poisons, real grenades, or fully automatic weapons. I could continue with things like napalm materials, mortars, and such things like that; but you need to make sure you consider the advantage of having the items versus it blowing up or killing you if an earthquake triggers a reaction or event you didn't want. I am not telling you should get illegal items; however, if you do get those items, you should have them in the deepest part of your haven and hidden, with the full knowledge that the items in that storage room is for emergency only in the mist of the crisis. Please do not take what I said as you should store your special weapon of mass destruction; no, you should not have anything that will destroy your safe haven or anything that helps out the crisis you are trying to survive.

It also behooves you to create a very well hidden storage area with fuel and items that can combust away from your safe haven. This storage area must be self sustaining when it comes to temperature control and all. If that is not possible, then its okay, but you will have to make sure it is well insulated so it can withstand the extreme heat and cold of the environment. You want fuel for vehicles you might have, generators, and also to help with burning possible corpses and unwanted vegetation or debris outside of your safe haven.

There is a very long list of supplies you need or should have in your safe haven, and I'm not going to go down a path which is different for every situations and group. But I will say that your list needs to address all the needs you currently have in respect to medical, sustenance, shelter, security, communication, education, consumables, maintenance, and fabrication which I will address now. Being knowledgeable as a machinist is a very valuable skill, especially if it is paired up with skills like AC/Heat, electrician, and a few more. Being able to fabricate parts is a skill that will save you a lot of problems in prepping, but it will help more in the long term goal of survival for a lifetime in your children's future. This skill allows for a more diverse ability to store and create supplies you need but can't afford the space for. Also note that it's easier to store items inside their packages which require assembly, if possible. Many preppers use vacuum seal plastic wrappers which are a must to have; to help your storage of supplies and other items you want protected from water, humidity, air, dust, bacteria, and fungus.

Remember that I said just about everything has an expiration date. I recommend five years of supplies, which is different than two-three years as most experts recommend. Some items can still be eaten or used after the expiration date, but I don't recommend it since you will be taking chances that the item is not contaminated. It is up to you, because items can still be contaminated even if they are not expired. There are organic or natural items which last longer and are better for you and the safe haven. There are also many creative ways to make things last longer, like coating food with specific other foods that help preserve foods like mineral oil or chocolate. There are also foods that can substitute for medical or hygiene needs. Amway, a general distribution company, has the best cleaning products by far which are mostly biodegradable, and can be used for many things to include skin care. Nature Sunshine has natural products which I highly recommend, like their toothpaste products. There is a problem though; the average expiration dates on these products are about two to three years. So, how do you get more than three years of supplies without breaking the expiration dates on most items? You will have to get creative, and understand also that if you store items at room temperature or very cold temperatures, many products outlast their recommended or labeled expiration dates by a year or more. You can also work on having an extensive hydroponic room and many plants can provide the Aloe Verde and essential herbs and natural remedies for many things you will need to survive. Baby shampoo is a must, plus your own personal shampoo is important, but don't get carried away with getting different and specific shampoos, or body wash for each person. Simple general supplies for everyone makes things easier to store and keep from having envy develop in the children later. Special items like dandruff control and shampoo for fighting lice and ticks are very important as many other items you can brainstorm. I hope you can see that supplies is not just a few different items, or a handful of one item and think that will last two years or meet all needs.

The subject matter experts or adults assigned to take on their assignments in medical, security, communications, and others must make a list of the items needed and why. A compiled list (draft) is exposed to the group, and all the members address the items as to challenge things they think should not be on the list and why. If there is such a challenge the reason needs to be provided, but also a solution. If there is no solution, then the reason which the expert stated for having it on the list should stand. Quantities of the items are also something everyone needs to come into agreement on. All of this sounds time consuming and requires every adult to pay attention to detail and listen to all the listed items. And it is time consuming; however, this is a prepping phase event that needs to be done in order to better see what requirements are needed as far as space and construction is concerned. The leaders should be the final determination if the item is a valid item, or if items are missing. I am not talking about food as the focus of this supply list. I am talking about all consumables, medical, weapons, furniture, electronics, communication equipment, items like batteries, POL, machinery, vehicles, appliances, spare parts, construction equipment, and security equipment. This will set your goal as far as what you want to stock up on. Hopefully, yours and other's hording will take care of many items on the list. Construction materials are last and should be talked about once the entire supply list has been established, because you will use construction material for your safe haven; any extra materials will be used for special projects and the rest in the end will be stored. So you really don't know the exact amount of materials you will have left over to store. This is another reason you should anticipate an outside storage area for things that become extra.

Many people think MREs are great, and they are if you are out in the wild with nothing else to eat and are doing things. The US Army MRE has a lot of calories which is in fact not good for you if all you are doing is sitting in your safe haven all day long. If you are busting your behind all day using up those calories, more power to you, because a MRE is probably what you need or can eat. There are however, meal replacement mixes which have all the nutrition and calories you need sitting or not. Most are Soy based products, but there are good natural ones that won't harm you as some people suggest they give you health issues, in women in particular. The rule of thumb is you don't want to use one type of meal replacement, MRE, or artificial supplements on a daily basis. You want real natural food, cold and hot; but your supplies have to cover all the grounds. If your plants/harvest or livestock/fish are not producing enough in a lacking time spell, you need to have other supplies to cover the gap.

One last note on supplies. People have habits which do consume more supplies than normal or needlessly because they do not want to change their habits. It is important that these people change their ways; however, you need to stock up and build your safe haven storage areas in accordance to abusive supply usage. There are many movies, books, and shows where people fight over the last few bottles of water or food because someone didn't want to follow the rations control procedures. If you think about what I said about routines earlier, you should be able to get the group to routinely conserve supplies by getting everyone involved. If everyone does their part, you won't be fighting over who gets the second to last straw.

Chapter Eleven

* I * I *

PREPPING AND EXECUTING ARE DIFFERENT ANIMALS

Many people have this idea that preparing and executing are the same things when it comes to survival. You see this in the show Doomsday Preppers, but they are completely different. In the show, the preppers more than half of the time are in the execution mode. They practice their escape routes, test out their equipment, run through drills, or create things to help in survival; but what most of them do not do is separate the preparation and the execution of their plans. Half of the people also do not have a comprehensive evaluation or feedback type of system (like AARs) which causes them to adapt to changes to the overall plan, which they don't do until they see their battle drills don't work or else they deny the gaps and call it a victory.

An example of what I'm referring to is: imagine you want to build a house from the ground up. You do the preparation and you get the money and find the location. You buy the land and make the foundation. You put up the frame and you get the roof on top so that when the rain falls, it won't destroy the wood frames and floors. But before you complete the walls, you bring in furniture and realize you need power, so you get the electrical working and start living inside the house without walls thinking you will get to it as days pass by. Before you know it the weather changes and you are watching television huddled next to a space heater under a completed roof and a house with incomplete walls and lack insulation.

Well this is what many preppers do. They get their foot in the door and skip construction steps. I am not saying you shouldn't learn hand to hand or practice at the firing range, and not build your bunker at the same time. No, there are things which you should do at the same time to maximize your time and efforts. What I am saying is you shouldn't create an elaborate hydroponic farm in a spare room and then realize you need to spend more money and time in moving it to an underground bunker. Build the bunker first and if uncertain as to the outcome, make an experimental small farm to see if it will work, and you can incorporate it into the bunker plan with a fully operational farm with room for expansion.

There are people who learn <u>extensive</u> martial arts skills which would be useful in close quarters, but any bum off the block can shoot you from a distance and not know a thing about hand to hand combat. The prepping part should have been asking the question of why would you need to get in close quarters in the first place. Then you look at what are the most likely situations with respect to what is really happening in the world to what you need to do. In this example, you would look at prioritizing what you should train on first, hand to hand being one that can be started half way through your creating a safe haven. If you have the time or access to expert trainers, go for it. There is also the aspect of gaining confidence which is something you need in the prepping phase. Now you should focus on close weapon systems besides hand to hand proficiencies. I say this because it is easier to learn pistols, than hand to hand. It will also be more likely that you will be faced with a close quarter situation which can be taken care of by using a pistol or non lethal weapons like a TASER. Along with pistol training, there are really good trainers who teach you what to do with pistols in a hand to hand situation; therefore, giving you an option with your pistol in a hand to hand situation. You always see in the movies where the Soldiers fix bayonets. Most Soldiers are not hand to hand experts and you don't need to be either from the start. They train Combatives and fighting with a bayonet, but their training is not to the level that top degree martial artists are, nor do they need to be.

The heavy battle drills and expert training should be done in the execution phase, which is when you have at least 75% of your safe haven structure completed. All of this prioritizing is handled by the leader with group input. If for some reason the leader is not considering the group's input; he/she should give reasons as to why the input the group is suggesting is not in the best interest of the group's survival. This will make the leader have to listen to the group and not just take the reins thinking he/she is always right or knows better. There can be many things go back and forth with heated discussions if there are many different goals and ideas in the group; but hopefully you will have a flexible group and can separate what you think should be part of prepping, executing, or both.

Prepping and executing for the adults is different than the children. Adults might get the children involved with prepping early on but, you need to look at prepping as if the execution phase has already started for the young children. The younger kids will follow your lead only because you are the authority and they must follow you. They will see everything you do as in the execution phase, not prepping. You don't need to waste your time explaining the difference; all you need to do is prep and focus on getting to the real execution phase where you are improving your skills and adding on top of the goals you have already achieved with supplies, structures, and capabilities. When they mature and really understand what you are doing or have been doing, then they can see the difference and focus on what is needed, not just putting the entire experience in their minds and hearts as one big survival lifestyle they inherited.

In Doomsday Castle the group is comprised of all adults (an adult is a 16 yr old or older that is mature to the point of understanding and accepting the reality of the doomsday event). You can tell that it is dramatized because of the show; but I like to use Doomsday Castle as an example because you can see that the father took a logical approach to training, constructing, and maturing his children to accomplish something they never did as an entire family. The problem is they basically moved into a half constructed castle sort of speaking and started the training without really learning how to can food, and other skills before you start battle drills. In essence due to show dramatization and to show the cool stuff, they started in the execution phase and dabbed into the prepping phase to give the members skills they should have already known before conducting battle drills. Granted, the family overall knew a lot about survival and prepping which gave them the confidence they had on the show; but like I said, it was more of a drama show, and not a learning step by step demonstration. If the children were hosting the show and go over the things they did and why, then it might have been like a DYI network show in construction and survival, but it was geared as a reality show, not completely focused on being educational. The construction of the castle was interesting, except that it would have been more beneficial to the family members and the audience if they had showed how the blocks of stone were piled or made. I'm sure a construction crew came in there at one point, but at least people can see that some things will require a lot of peoples' involvement. It is possible to do the foundation of your underground or above surface structure with a handful of people, but that is only because the people are very knowledgeable and experienced in pouring concrete, reinforcing, and the like.

Okay, back to prepping and execution. If you have limited resources, you should focus on getting more resources. You can't really go into an execution phase until you have an established safe haven; whether it is your own property or another prepper's property, you are at a standstill without a place. Now, there is a possible option that you don't need a plot of land or house; however, this only works for specific doomsday events, like an economic breakdown or food crisis. You can stock food and the like where ever you are and get a vehicle or trailer to tow the items to a safe area you found. It might be in the middle of the woods as if you were going camping, and live out of a trailer or tent/vehicle configuration. Granted, chances are you will probably need to hunt for food, plants, and find a water source. This is where you can see that linking up with other preppers is important. Three preppers with limited resources make one group with many resources. So in this case you can see that prepping is not execution. Prepping in this case is finding other preppers and coming up with a plan to find a location they can meet at and establish a safe haven. One prepper can focus on construction items, the other on electronic systems, and the other on weapons. All three can stock food and other items as discussed earlier. When they go to the site and actually practice the establishment of the safe haven and make caches, or whatever their plan entails, then that is considered the execution phase.

If the plan is on creating an underground safe haven, they can all try to buy the land as a group, but I would only suggest doing this if there are limited resources and you can trust your partners with your life. In other words, all of you should be seeing each other as true lifelong friends before you consider this; not boyfriend or girlfriend like I said earlier. Even in spouse relationships, there can be a lot of friction if the marriage is young. This buying of land is not a prepping phase event; this buying of land is a commitment and should be seen as an execution part of the plan. If, in worst case scenario, there is a falling out, there should be an out for the owners, and one or two people can sell his/her portion of the claim to the land to the others. There is legal stuff involved with this and research should be done and measures taken to cover a falling out, _even if_ _everyone is sure_ they will not have a falling out with their true friends.

The military has a method of training which is broken down into three steps. The crawl, walk, and run steps to conducting or exercising a plan or action. These steps are used extensively in rehearsals. Of course a rehearsal is associated with an execution of a mission or actions at the objective, is what some people would call it. But I want you to look at the overall prepping as a crawl and walk, and execution as a run step. In the crawl step, people are literally walked through each step of an action. I will use a simple action like mowing the grass as an example. I would tell you what you will be doing using diagrams, show you the mower; teach you all you need to know about how to operate it and all. Then we will go through the procedures of checking the mower, preparing it for operation and then turn it on. Once that is done, I would show you how to clear the lawn for items which would damage the mower, then I would show you how to push the mower over the grass in an efficient pattern so the grass is cut evenly and has a nice appearance. You would tell me and demonstrate to me what I have just taught you and push the mower to indicate you understand the pattern and all. In the walk stage or step, you would go through all the steps in putting the mower into operation and cut a portion of the lawn to show your competency. Once, you have mastered the walk step, you will go through the run step and cut the entire lawn without me asking you questions or giving you any guidance. In the end you and I will have an AAR and evaluate the entire process from crawl up to run/ start to finish of training.

This step process should be done the same way for all your training and rehearsals; but for the purpose of separating prepping from execution, you should use these steps as follows:

Your crawl step should be research, meeting with or recruiting preppers, identifying the event(s) you want to survive, making goals, gathering information, selecting a location(s), and gathering resources or increasing resources.

Your walk step should be all of the crawl steps plus: getting at least a six month stock of <u>all supplies if possible</u>, obtaining land, and construction started up to full completion if possible; gather specific resources like solar, air filtration systems, and the like. Additional preppers are indoctrinated with the crawl objectives and trained once there is a clear assertion that the new preppers are in fact on board with the original groups' goals (remember goals, not necessarily beliefs, which I talked about earlier in the book). Survival or mission essential vehicles are obtained or in the process of being obtained.

The run step is basically the execution phase in the big picture of the plan for survival. If there is any construction for the main structure to be done, it should be completed to the point of having all of the electrical, air, water, and waste disposal systems fully operational. The only major construction you can still be doing is fortifying the structure to withstand extreme heat or cold, and heavy munitions. Your outside storage should be completed <u>or started on</u>. Your role/position in the group dynamics should be focused on battle drills, your basic weapon skills, hand to hand, and daily routines should already have been firmly established. If you go deeper into these skills, that is great and should be done on a continual basis, especially for the children and upcoming leaders.

Even though the U.S. is not in a state of war like in World War I and II; many people stand on the idea that Americans and others are at war against terrorism. I agree that this is a real war, and I recommend you look at your prepping and execution phases as a war. The military and many civilians have been in the business of waging war throughout history. Waging war is in the form of creating all sort of logistics, performing training, and using the war machines against the enemy. Your preparation for a war will consist of very logical steps and you must decide how you must go about doing it. I have suggested a method of performing tasks using the crawl, walk, and run method. I have also suggested you look at the prep and execution phases as separate animals. However, this is war and you should research other options in how you track and work on completing goals. My suggestion is you use military methods as opposed to civilian law enforcement or even streetwise methods. If you analyze the things I recommend, most of it is based on military methods or ways of looking at things. One thing the military teaches you is to always keep your options open, don't recreate the wheel, and be aggressive without doubting. If you split the prep and execution phases, then you should do well to achieve your goals.

Chapter Twelve

* I * I *

PUBLIC RELATIONS POLICY

The best friend you can have should be your spouse, and anyone beyond that is either family, true friends, unknowns, or enemies. The relationships you build is not what will determine who you will trust with your safe haven or prevent from compromising your survival chances. The adaptability of that relationship is what will determine if you can trust someone with your secrets. People change their minds and hearts based on things that they fear, hurt them, or think to be good for them. This topic goes into the realm of personalities, and I could write an entire book on the subject, but like I said earlier, people tend to change because they want to or they are forced to change. An example is if someone loves eating sugar at very high levels or drink alcohol every weekend, they will not want to stop over eating sugar, or drinking. But if they become diabetic, they will either try to substitute the sugar with chemicals or reduce sugar intake and stop drinking out of a desire to do what is right to live, and not feel like they are being forced to change because someone told them to change. This example is simplistic, but like most prisoners who come out of prison, they are forced to change their behaviors while in prison, but once they get out, they tend to return to their old habits, not because they have to, but because they never really wanted to change their ways. However, many also don't go back into crime because they have learned their lesson, sort of speaking, hence they were forced to change which is reinforced by the threat of prison time again or worse, like death, if they go back to their old ways. Your closest relationships must be based on acceptance, wanting to be in the relationship, and adapting to changes. Understanding your roles in the relationships will keep each other from betraying one another.

It is very unlikely that a relationship would be unbalanced in the family if everyone is in agreement with the survival lifestyle, but if there is, you should work on giving the members who want out, a way out. As for the members who are not family members, they should have been handpicked as I spoke about in the group dynamics. So what this chapter is mainly for is the relationships you come across with people and animals outside of your group.

I mentioned true friends and unknowns. True friends are people you strongly believe that will die for you. They are never judgmental, give you good advice, think the best of you, and trust you to do the same for them. They maybe in your group which is fantastic, but I'm not talking about those true friends. I'm talking about those who are not in your survival lifestyle and when things hit the fan, they might come to you for help. Or they might stand side by side with you through your prepping period and not want to or are unable to commit themselves to prepping. This may possibly bring you to one of several major dilemmas you will face when things do hit the fan and they somehow were able to get to the door step of your safe haven property or perimeter and want to join you. If you were paying attention and know that they were not able to join your group because of something sort of out of their control, like they were far away nursing a loved one, or have to care for their family and won't leave a good paying job. Well this is a good thing, because if you were paying attention you would have made some room for them to join.

The situation is different when you get the Twilight Zone crowd of friends who you love or like as good people, but were totally against you prepping and now they are begging you to take them under your wing. So do you shoot them before they damage your safe haven systems or structure; or leave them stranded out in the open subject to the crisis; or do they come in and your survival chances are greatly reduced because your safe haven is overcrowded or occupied by completely different ideologies. There was a Twilight Zone episode where the neighbors came in the house and broke down the shelter door. The nuclear threat was pronounced a false alarm, but the old man who owned the shelter saw the true character of his supposed friends/neighbors who said kind words on his birthday party that same day. But cursed and threaten to kill him if he didn't let them in the shelter which was built for three people.

Jaime Mera

There is a reason why you want your safe haven to be in the middle of nowhere or at least away from most people. If you have to make that tough call and shoot at your friends or neighbors you know, then so be it. Survival can be looked at not whether all of you can survive, but are you willing to sacrifice your family to give someone else some mercy and comfort; and in the end all of you will die or most of you will die because having too many people to feed and live with will cause your survival chances to decrease dramatically. If your doomsday situation is against a super volcano, meteor, or biblical events, then having too many people in your safe haven is not what you want and will make everyone suffer in the end. Not that you might not have enough food or air issues; no, you will have space issues which can and will cause social issues which can get people killed. If someone is killed (not dies from disease, accident, or old age) and there is no way of disposing the body, then you have a major health and emotional issue within the adults and children. It is a call you will have to make, based on can you or the group afford to take on an unexpected new member(s) into your now execution phase in a crisis.

Now, it is easier to repel hostiles or unknowns by warning, threatening, and shooting/attacking them so they stay away from your safe haven without for the most part feeling guilty of doing something wrong. In an economic collapse, this will be harder, because if I were a person who never thought about prepping, but know military/weapons skills I could get a group of desperate people and go hunting. Hunting for targets that are easy to take and give me the best resources I am looking for. My public relations policy will be invade, shoot if shot at or threaten, and whatever I do, I will not engage defenseless people; in particular elderly, females and children. Not that I am chevalier, but in a strategic stand point I don't need to have a feared reputation, only a respected reputation. If there is some form of law and order, law enforcement will not be as deadly on me and my group, and in fact might simply warn us if we are doing something that is stirring up more problems. The idea is to let law enforcement see that you only defending yourself and fight for others. So if you help the community, it might save your group in the long run. There is this idea that martial law will cause people to have to give up their weapons to the law and order in the area; and this might happen; which is why my group will not be telecasting our presence. The game of hiding is how people will come to you, and how you should be moving around if you have to.

The rule of thumb in a crisis that leaves many people roaming around is to keep people at a distance, stay put with high security, make your safe haven a hard target to attack or molest, and lastly; never assume people are friendly, even if you know one friend out of many in a group. Unknowns are probably the most dangerous people you can run into. Enemies are easy to identify, especially when they have ignored your warning signs, or warning shots and continue to advance or sneak into range of their weapon systems. The people that you should worry about are those that seem to be lost and carrying weapons, or even scouts without weapons. It is not likely, but females could be useful scouts to find other people and safe havens/resources. If you get a woman to scout for your group, make sure she has non-lethal means of defense, and that you can over watch her every step. If a man is the scout, that is fine, but chances are that a man will be shot without warning before a woman will. A female carrying mace or a Taser, is not suspicious; but a man carrying those weapons is suspicious. Now, what are these scouts doing? They are assessing who is out there, what firepower is in the group, what resources are available, and details on terrain and possible traps.

Your scouts or their scouts will take intelligence on your enemy, or you to the enemy, and there you go. You will have an idea of whether it will be worth your or their effort to conduct hostile actions or try to make a peace treaty. If the scout method is not to your liking, then I highly recommend you work on sniper skills. Your ability to move into an area and gather information without anyone knowing it will save you a lot of pain and lethal injuries. Best case scenario would be to get a six team sniper group which is twelve people. It might not be as pretty as a military six team sniper force; but for what you want, it will be your method of gathering information. You will determine where to go and what resources to tap into, whether it is people with resources, or resources in the open – like a field of apple trees.

I bring this six team sniper force as an example, because you in your safe haven need to consider that unknowns are in the form of scouts and snipers. If you run into or see them, the best initial weapon against these tactics is to not have information shown to them. Hence, and underground complex is not visible by snipers or recon teams. A well rehearsed personnel processing system will keep a scout from seeing numbers of people, details of weapon systems, exact locations of where things are stored, and communication systems. Blindfolding a unknown scout will help, but even then they can see what you have if you let them inside your main complex. You should have a room that is like a holding cell, except no bars. Aide can be given to the person or people and released later, but it is this period of time you need to determine if they are scouts, really in need, or have other ulterior motives. If you have a processing plan, this is what you should consider. If you don't have a processing plan, but more of a "move on or be shot type of policy", make sure you have video so you can put that person or people in a database in case they return later, and also show all the members in the group who this person is in case they run into him/her or them in the future.

If you follow the recommendation of putting your safe haven in the middle of nowhere, then you should not have to worry about your public relations policy, because it will be simple. People that are in your area are either hostile or completely lost. If they have an organized backpack and seem like they know what they are doing, then it is your call to treat them as hostiles or not. In the end, unknowns can become a life saver, life taker, or just passing by without any problems.

Animals are different issues. Chances are that you will not have a stray dog or cat in most places, but if you do; do not let your emotions dictate whether you take the animal in or not. If you are not prepared to care for an animal completely, this includes able to give the animal all medical shots, other medical treatment, food, and domestic training; then I suggest you let the animal move on, or possibly feed the animal and let it live outside by itself. This can cause problems which can lead to you allowing the animal to enter your safe haven and hence you have another mouth to feed who might spread disease, or cause tick, lice, or waste disposal issues. Other animals should not be considered as friendly. Rodents are and always should be considered hostile; not food. People will argue that rodents have protein and are of value for food. The amount of protein in one rodent is not worth the effort and does not outweigh the potential of the rodent transferring disease and poisons into the safe haven or inside your group members. You might have this idea that cooking a rodent will kill disease or get rid of harmful things, but it isn't a guarantee. The major problem is that exposure to the rodent before it is cooked is the point of transfer of the contamination to you or a loved one. Fumes from rat droppings or dead rats can kill you in particular environments. My point is your enemy are animals; in particular, all rodents. Dogs and livestock like donkeys, horses, or food producing animals are and should be considered friendly. In a pandemic, all outside animals are enemies.

One last note. There are preppers out there that want to roam the lands to help people. I strongly recommend these preppers band together and know other preppers that are not roaming, but can be used as stops in case the roamers need a place to stay for a while or to get fuel, supplies, repair equipment, or medical attention. The problem with this is you might be inviting hostiles into your location or group. This is where you have to take a leap of faith and trust that the network of preppers you have associated with and plan to mix with are ones you trust (I didn't say can trust, I mean really do trust). Mobile preppers have it hardest since they don't have military convoy vehicles, and some are alone. One to three vehicles are easy targets. A convoy of five or more light to heavy vehicles are hard targets. These are vehicles on the move; and when in a stationary position have very good security. Almost all-terrain vehicles are usually hard targets, but terrain will dictate if they are easy targets in an ambush. Vehicle mounted weapons are devastating. Reacting and identifying ambushes or raids are the skills that will keep you from being killed or severely injured in a vehicle or on foot.

Roamers can have the advantage of caches of fuel and supplies if a stationary prepper is helping them by safe guarding the cache. In return the stationary prepper can get supplies from the roamer who is on the lookout for additional resources. Good communication is key to making everything work while on the move; and those people that are on the move should be visible to you in your safe haven. If they are not, there is a chance you will fire upon a roamer who is trying to help people. Your policy with dealing with other people and animals outside of your group will determine if you survive, go to jail, kill an innocent, or save other peoples' lives.

The intent on roaming is usually a noble one, but if you don't have the support, resources, and skills to get into and get out of fire fights, then you shouldn't be roaming.

If you happen to be one of those people who were prepping or started to prep before the crisis hit the fan, my suggestion is for you to search for a group that can take you in. Once again, knowing other preppers will help in this. Hopefully, you would have learned enough to keep you alive and be an asset for a group you do join forces with.

Chapter Thirteen

* I * I *

KNOW YOUR LIMITATIONS AND STRENGTHS

G etting an evaluation of your strengths and limitations is something you should never pass on. The people in Doomsday Preppers got an almost free evaluation, maybe not as extensive as I or others would provide, but it was an evaluation conducted and influenced by television producers, experts, and melodrama provided by some of the preppers themselves. The scenario of a pandemic as a doomsday crisis is real, and I will cover this in the next chapter which deals with biblical events in Revelations. However, in this chapter I will use an example of preppers who think they can fight a pandemic in their mansion home which was very large, three stories high. The experts suggested they move to a shelter and be separated from people for at least eight weeks. The leader disagreed and said they could fight a pandemic in their home; but it seemed that the leader simply didn't like the idea because being in a self-contained bunker was too much or not something the leader wanted to struggle to achieve. This is survival and liking something is not an option when it comes to survival because you want to enjoy the looks of your surroundings. There are decision points where you must see your limitations and strengths and adjust. The prepper who was unwilling to move to a self-contained environment didn't want to leave the mansion they lived in. The decision to stay and trust that the ventilation of the home was going to keep things safe was based on lack of facts and lack of seeing limitations. The leader was knowledgeable in quarantining people and details about making a <u>room</u> germ free sort of speaking, but they don't know how to prevent the germs, bacteria, or disease from entering <u>the home</u>. If you look at the black plague in history, you can see that the disease changed its form of transmission. Once the plague went airborne, there was almost no safe place. Burning fires helped but my point is, the doomsday scenarios that will be fatal to billions of people must be fought with complete self-containment.

The strengths you see as strengths just might be that, but there are always limitations. You will always have both and you must address both. One strength that all people have is the desire to survive. Your adaptability to survive is the first strength you must acknowledge and make better. Some people have rules like in the TV show NCIS where two is one and one is none, or always carry a knife. A knife is a multipurpose tool and weapon; just like having one plan is like having no plan. Backups are a rule of life, and it is easy to see in physical objects like a knife and pistol that they in themselves are backups for you; but it is hard to see when you look at a person's mindset. Confidence or belief that you can solve problems or know solutions for surviving is your strength; never a limitation. Basically, your adaptive ability or strength is based on your knowledge and ability to solve problems. This is where the leadership, group training, and most importantly your librarian/knowledge database is where you will get this confidence. There is a television show called, "Owner's Manual", which has two guys, one using common sense and intuition while the other uses nothing but an owner's manual to basically do anything with an owner's manual. These guys figured out how to run a locomotive, dune buggies, fly planes, and a lot more; without having to be certified or get training by experts.

I'm not saying these guys are the average people who can figure out how to fly a DC-10 jetliner with an owner's manual; but maybe yes. These guys are very adaptive and know a lot from reading and hands on with things similar to just going around learning things like a handyman in construction. For most things you can just read and learn it. There are skills which require actual skills with manipulation of tools, or body parts. But, for the most part your group is only as proficient as they want to be. It is easier for some people to learn by someone teaching them and showing them, but that is where the librarian can help. Owner's manuals and step by step instructions are key in being adaptive to just about any situation. Having said all this, teachers or subject matter experts are always a strength. If someone can't teach very well, than that is limitation; however, the strength is still on the expert's side. Find a way to teach the person how to teach.

I can go on with strengths and you should evaluate your strengths based on whether they give you an advantage in survival as compared to what your goals are for surviving the event you are looking at, and how that strength can help you. If it is a limitation, then it is because you have evaluated whatever it is to keep you from making the goal(s) for your survival plan.

A limitation which is usually a given is not having enough resources; but you need to keep certain limitations in a corner that almost never go away, Limited resources is one. There is no real value in stating the obvious, except if the limited resources is so bad that you can't get started. Usually this is because you are in the middle of a legal battle or a personal crisis and simply can't devote time, money, or anything into your survival plan. There are many things that require or could cost a lot of time, money, or attention. An example is Dragon Skin which is a type of cloth armor used as body armor for people, livestock, or objects. There are different levels of Dragon Skin and if you research it you will find a variety of information for and against using the material. But it is expensive if you plan on using a lot of the skin for children, backpacks, body armor, and livestock.

The rule of thumb for evaluating your limitations is to first evaluate what you think is a limitation, and then get input and listen to what everyone in the group thinks is a limitation. A solution can be brainstormed so hopefully you will have only a dozen or less limitations listed. As you go through your prepping and execution measures, you will hopefully reduce the list, and chances are you will also add to the list as you adjust on a weekly or monthly basis.

Depending on your group dynamics, your limitation maybe the fact that you and your spouse are the only adults in the group. This is a limitation, because your goals cannot all be performed to standard without more adults. I told you the ideal group is six-twelve adults for many reasons, and one is so you don't have this limitation. At the same time, the group must work effectively together, otherwise it is like having three couples all working separately and are limiting you in achieving your goals. You can hopefully see that you need to develop your own list of strengths and limitations because it depends on what you know, what you have, and the opposite of what you are lacking.

I must address the safe haven construction in this section of the book. Many people think that an underground safe haven is too much or not needed at all. The above surface safe haven which is self-contained is also considered too much by many preppers. The chances of an economic breakdown are very slim. The US economy is too strong, not in the sense of the dollar, but in the people. The great depression didn't come to a doomsday event; the EMP solar flare that knocked out electrical telegraphs in 1859 was not the end of the world. There might be riots, but that is not the end of the country or life. Food is there and will be there; martial law and the military will keep law and order. Preppers will be ready for these events; but what I said earlier was this book is for worst case scenarios. What will cause a severe economic collapse is a super volcano, super meteor, or biblical plagues (pandemics and worse). It will make it so people will not have access to food, like farms or backyards, and livestock which will be killed. A self-contained environment is the only way to avoid these issues. Now that you understand this; not having a self-contained environment is a limitation. Your goals will not be achievable to many degrees. You can disagree with this, but in the end you cannot call having a safe haven without a way to keep outside toxins from getting in as a strength. The Doomsday Castle is impressive, but in a super volcano they can probably survive for a while if not in the fallout zone, but in time people who are desperate will overrun it, or the pandemic caused by the volcano will destroy their efforts because it is not a self-contained environment. Another note, which I wanted to address is your underground safe haven should be five meters deep to the top of the ceiling of the rooms, not the floor. There will be several areas that are at ground level, but for the most part everything else should have five meters or more for protection. It is very unlikely, but the military has munitions designed to penetrate ground and completely whip out your bunker. What the average hostile will have is your high powered rifles, grenade launchers, and in a likely scenario a mortar. If they are mobile they could have these types of weapons. An above surface target will be susceptible to these damaging weapons; unlike an underground complex. Note: I recommended a safe haven inside the side of a mountain. This eliminates indirect weapons from hitting you and most direct weapons being high in the mountain. If you decide to go deeper than ten meters, that is fine; but know it will cost more the deeper you go, and the reward will not be much better than ten meters. If the enemy uses something that can penetrate ten meters of ground, it will do major damage to anything within fifty meters. Not that you should go sixty meters or more on a flat surface or hill, but on the side of a mountain you want to go as deep as possible sideways if you have the resources. Make a massive complex if you can. Having too much space is always an advantage and strength. More space means more air and room for expansion of hydroponics and many other things your group decides to utilize that space for.

So how do all of these identified strengths and limitations play out in your survival? You breakdown a list of strengths and limitations to go with each phase of your prepping. Since they are linked to your goals, you should have four lists, one list to go with your prepping short term goals, and one list for your long term goals; then one list for your short term goals in your execution phase, and one list for your long term goals. Update the list as needed during all the phases, weekly if you have to; however, if you are editing a strength or limitation too much, then you probably have not really identified it correctly. Many of the items on the lists might not change, but chances are there will be differences and there should be differences because you are supposed to be constantly trying to eliminate limitations; and capitalize and add strengths as the group increases in skills, knowledge, and resources.

I talked about dogs earlier, and will use this as an example. If you have a German Sheppard which is considered one of the top ten dogs to have when it comes to security, they can be trained to attack, sniff out, assist a person, and all the other things dogs normally do. However, if the dog is not trained or is a wanderer you picked up, it could attack your children and all the negative things associated with dogs. This is not a limitation, but having a very well trained dog is a strength. A hostile or problematic dog becomes a limitation when one of your goal is to maintain a hidden and quiet safe haven and the dog is constantly barking at everything; because your goal cannot be achieved as long as that limitation is present. Do not get too general by saying that if the untrained dog attacks my baby infant it is a limitation because one of my goals was to keep the children safe so they can procreate later. No, that is not a limitation; that is a problem you must fix immediately as if it were a hostile force coming into the middle of your prepping or execution phase. Anything that presents an immediate lethal or serious threat to the group is not a limitation; it is a problem that needs fixing at that time. If you all have to put things down and devote the entire day or month to fix it, then do so. Other examples are people constantly not following safety procedures with firearms or while construction is going on and there are very heavy objects being moved, installed, or created.

A limitation in most cases are things like we don't have enough food to last five years (if your goal is to have five years worth), we only have six months worth. That is not an immediate threat since you do have food, but just not enough of it to last five years. Another limitation is you don't have a medical expert, but you do have access to doctors while your medical expert trains up or while you find another prepper who is a medical expert. In most cases, a limitation is something you lack, but in time it can be fixed. A strength on the other hand is something you already have, and need to exploit. If you don't exploit the strength, <u>then it is not a strength</u>; it is simply a skill or resource that is sitting doing nothing. You want to look at it that way, but when you write a list of the strengths, include them even if you have not exploited them. What you could do is annotate if the strength is being exploited, or when you plan on exploiting the strength. If you don't plan on exploiting the strength then why list it? It is a good question, but you never know if a situation will cause it to be exploited. An example of this is if you are a salesman and know how to influence a crowd of people, then when you meet other survivors, you might have to use this skill to get everyone on a survival mode instead of a self only mode. You wouldn't really need this skill in prepping since the people you are prepping with should be on a non-selfish survival mode. The librarian should be doing his/her job by noting that it could be a strength, but not used. This will help others later if the strength can be resurrected in case it is needed in the aftermath or in the execution phase.

If you link the strengths and limitations to the goals, you will find out that they will help you in prioritizing things. There are many ways of doing almost everything, and it is up to you and the group to decide how things will be prioritized; but following the goal setting recommendation along with placing the list of strengths and limitations will help you see the gaps in your overall plan and priority list.

Last note: there was a prepper who got an assessment on their security measures, and he was asking who are these experts? He said they didn't know what they are talking about. Well the experts suggested he get more cameras on the ground to give his group more early warning and situational awareness. His group consisted of his wife and seven children. The oldest being twelve if recall correctly. The parents were experts in weapons and taught survival skills for a living. The children were also very proficient in weapons and things like sniper tradecraft. Their safe haven area was in sparse desert with vegetation. Not out in the open, but there were no major mountains or very large hills. Here lies the problem. The parents believe that they and maybe five kids can pull 360 degree security 24/7 for months if not years without cameras and communication systems. Granted he might have had cameras or a communication system, but he didn't show it to the experts and he didn't reveal he had those assets. My point is that by just watching what was shown on the show Doomsday preppers, he didn't consider the limitations which two adults and a handful of very young children had in maintaining a very hard level of security. It takes a squad of 12 people their full attention to proficiently maintain 360 degree security 24/7 for many days if need be; but that is not good practice without early warning devices and a reliable communication system in place. The leader of the group was like you experts don't know what you are talking about; which in some cases I could understand having read book critics and also compare my expertise with supposed experts; but in this case I have to side with the experts. If the group was comprised of 12 adults and/or mature teenagers, then I could see that there might have been a security plan in place the experts may not have noticed or thought was ineffective. But that was not the case. I said before, children are force multipliers, not permanent adult replacements; until they themselves grow older of course. Being force multipliers does not mean they are limitations or strengths; however, the fact that security goals are being limited because they don't have a reliable 24/7 way of maintaining situational awareness in their surroundings is a limitation. What is critical is that people accept critics and really look at them/ with others, and analyze them to come up with an appropriate response, not an emotional response thinking you know it all or are right. I reassess myself sometimes when people tell me something contrary I believe to be true or false; and do my own research. Once I have come to a conclusion, I might not think the same way; but what I don't say is who are these people; they don't know what they are talking about. In the case of the leader responding to the experts, he should have said I have measures for security which are better than or as good as cameras or electronic early warning systems and everyone in my group has comms. Otherwise he could have also said "okay, we will work on that"; instead of "who are these experts and how did they come to this conclusion? You experts are so bogus."

The response of this intelligent person is an indication that limitations and strengths were not written down in any goals of a plan. Writing things down is a must for all plans, to include following what was written. Do not get into the habit of going with your experience or memory on how your plan is going and where it is taking you. If you do this, you have created a limitation on yourself, your group, and don't even know it.

Knowing your strengths and limitations is so important that I made a chapter on it. All of this might sound tedious and a lot of paperwork which needs to be monitored. And yes, it is tedious, but if you embed the information in your goals, like to the side - a strength and limitations list, you can always have the plan with goals on a wall to include the list of strengths and limitations and take your time to prep without missing something crucial. The crucial parts of the plan must address what you are trying to survive, how, why, when, who, and where. The <u>how</u> portion of the plan is basically where you have your specific goals and list of limitations and strengths.

Chapter Fourteen

* I * I *

BIBLICAL PROPORTIONS

Today is a day you must put science and religious beliefs to one side, but don't put them too far away from you. There is no way you cannot believe in a worldwide doomsday scenario without considering the Holy Bible. I do not mention other 'religious' sources that say the world will come to a sudden and horrifying end, simply because there is no other source that can top what God says will happen. The biblical events will not happen all of the sudden like in a blink of an eye, (which is misquoted by many people as are many other quotes from people who take things out of context) but they are definitely more horrifying than any other type of end which concludes with eternal suffering in a lake of fire for those denying Jesus.

I'm going to use the logical approach to this because many

people will try to argue with what I'm going to say about what the bibles says. I am fully aware of many interpretations and will address some in accordance to sound compliance to words, references, and context. I am also not going to point you to other books of the bible which say there will be wars and rumors of wars, etc; because what I want to show you is what you can imagine with a doomsday event that is laid out in one book and in sequence. The books of the bible are like chapters in a book; except the bible is broken down into Old Testament, chapters/books talking about the old covenant and before Jesus' ministry, and the New Testament are the books that tell of Jesus' ministry, salvation, and beyond.

The book of Revelation is the last book of the Bible and is 22 chapters long which is also the same number of letters in the Hebrew alphabet (nothing is by coincidence). When I address scripture I will use parenthesis with the name of the book, chapter is the first number and the verses are indicated after the colon. If you have read or seen a bible you should already know this. I made the biblical writings one font smaller so you can see when I am writing and not quoting even though I do use quotation marks and labels. I bolded the portions of scripture I want you to focus your attention to so you can relate it to a doomsday event or results.

I will start with Chapter 8, which talks about the trumpets which designated angles will blow in accordance to God's promptings.

First Trumpet

(Revelation 8: 6-8) "⁶And the seven angels who had the seven trumpets prepared themselves to sound them. ⁷The first sounded, and **there came hail and fire, mixed with blood, and they were thrown to the earth**; and **a third of the earth was burned up, and a third of the trees were burned up, and all the green grass was burned up.**"

You can take this hail and fire literally, or see it as a solar flare, small meteors, weapons like nuclear missiles, or something we have yet to see – like a particle beam weapon from a satellite platform in space. Notice that events resemble doomsday scenarios already and we have only talked about one event.

Second Trumpet

(Revelation 8: 8-9) "⁸The second angel sounded, and **something like a great mountain burning with fire was thrown into the sea; and a third of the sea became blood, ⁹and a third of the creatures** which were in the sea and had life, **died**; and **a third of the ships were destroyed**."

This depiction is very much a description of how a large meteor would look like and the damaging results it would have on an ocean. It could also be a nuclear explosion in the ocean. Remember that a modern nuclear bomb is massively more powerful than the test bombs used in the Cold War and on Japan. If a submarine with nuclear warheads on it goes off it will kill a third of all sea creatures and probably take out a third of all ships in the world since they will be in the path of a massive global tsunami.

Third Trumpet

(Revelation 8: 10-11) "¹⁰The third angel sounded, and **a great star fell from heaven, burning like a torch, and it fell on a third of the rivers and on the springs of waters.** ¹¹The name of the star is called Wormwood; and **a third of the waters became wormwood**, and **many men died** from the waters, because they were **made bitter.**"

This I believe is a meteor which hits ground, not water. The reference to wormwood and bitter water means that the Wormwood/meteor contaminated the water in rivers and mountains which made them poisonous; hence the meaning of bitter. This is one reason I recommend you have several sources of water even if you are tapping deep in a mountain water source, and why it is important to have a very, very effective water filtration system.

Fourth Trumpet

(Revelation 8: 12-13) "¹²The fourth angel sounded, and **a third of the sun and a third of the moon and a third of the stars were struck, so that a third of them would be darkened and the day would not shine for a third of it, and the night in the same way.** ¹³Then I looked, and I heard an eagle flying in mid heaven, saying with a loud voice, "Woe, woe, woe to those who dwell on the earth, because of the remaining blasts of the trumpet of the three angels who are about to sound!"

The third of this and that is the debris/fallout that will be caused by a super volcano, meteor, or nuclear explosions and will darken the sky 24/7. Keep in mind that people will go into the ground by this time (underground) and so as you read on in the bible the fifth trumpet explains sort of what will happen. There are several interpretations; like people will have super genes or medical advancements that will keep people alive. But if you read below you will see that things get worse, not in death but in suffering.

Fifth Trumpet

(Revelation 9: 1-6) "¹Then the fifth angel sounded, and I saw a star from heaven which had fallen to the earth; and the key of the bottomless pit was given to him. ²He opened the bottomless pit, and smoke went up out of the pit, like the smoke of a great furnace; and the sun and the air were darkened by the smoke of the pit. **³Then out of the smoke came locusts upon the earth,** and power was given them, as the scorpions of the earth have power. **⁴They were told not to hurt the grass of the earth, nor any green thing, nor any tree, but only the men who do not have the seal of God on their foreheads**. ⁵And they were not permitted to kill anyone, but to torment for five months; and their torment was like the torment of a scorpion when it stings a man. ⁶And **in those days men will seek death and will not find it; they will long to die, and death flees from them**."

The bible mentions that people will ask the mountains to fall on them so they would die and the suffering would end, but that won't happen, and people won't be able to kill themselves. Remember what I told you about survival. You want to survive with comfort and with a sense of hope; otherwise you might as well go out into ground zero before you suffer a painful and slow death once the big one hits the fan.

Sixth Trumpet

(Revelation 9: 17-18) ¹⁷And this is how I saw in the vision the horses and those who sat on them: the riders had breastplates the color of fire and of hyacinth and of brimstone; and the heads of the horses are like the heads of lions; and out of their mouths proceed fire and smoke and brimstone. [18]**A third of mankind was killed by these three plagues, by the fire and the smoke and the brimstone** which proceeded out of their mouths.

If you look at your math and fractions you will have logically concluded that one third of this and that is actually more than one half of all humans on Earth, because even a third is a lot of deaths, more than two billion with our current population. If you add all the wars put together from as far back as you wish since before Egypt, you will not be able to come up to one billion deaths of all deaths in wars. This is one reason I don't focus on wars and deaths associated with them as a sign for a doomsday event. I would look for deaths from massive pandemics, meteors, super volcanoes, and the like. So in this case I believe it is a massive pandemic due to the effects of the meteors, nuclear devastation, or super volcano. A polar shift <u>might</u> do the same thing, but it is unlikely it will happen from a scientific and biblical perspective. People in this age are applauded to the death of 3,000 lives all at one time, like the 9/11 total deaths in the US, which is very tragic and was completely conducted out of hatred and evil, not a desire for love of life. But to put things into perspective: there has been many more than 3,000 deaths at one time in history, more than a few times too. One time meaning one day or part of a day. The difference with 3,000 people dying at the foot of Mount Sinai, or more in Nagasaki or Hiroshima than with 9/11 is that we (most of us in this world) didn't see it or experience the deaths of so many in Jordan or Japan - live or on replay on television in high definition. So when you witness the death of so many people, you have a despairing sense that death or the end of time is coming. Unfortunately, we have yet to see the billions of people that will die in a very short period of time. You must be ready to accept and cope with the fact that many people will die either a very quick or long and agonizing death.

(Revelation 9: 20-21) "**²⁰The rest of mankind, who were not killed by these plagues, did not repent** of the works of their hands, so as not to worship demons, and the idols of gold and of silver and of brass and of stone and of wood, which can neither see nor hear nor walk; ²¹and they did not repent of their murders nor of their sorceries nor of their immorality nor of their thefts.

Repent meaning changing their minds or beliefs in looking at God but instead continue in their evil ways/sinful living without Jesus' saving grace. Apparently the fact that everyone else is dying around them is not enough for people to look at God for salvation. I think people will think that since they are alive, then they can't die; which is what will kill them in the end. If you follow my recommendations for a self-contained safe haven in the mountain regions I stated; chances are you will survive until the seventh trumpet; however, you will not survive the bowls of wrath and God's final ultimatum to those who have not accepted Jesus/God as their savior. Well, I know you think that I maybe pushing my beliefs on you. No, I'm telling you what the bible says. You can be a Christian and not believe it, but does it matter? I told you earlier, that if the bible is false, then you have nothing to worry about except maybe science, nature, or coming back in a different life form as an abused animal or tortured human in poverty or war, or food for other animals. But, if it is true, then you should mediate on these things as you are in your safe haven seeing things unfold as it says it will in the bible.

The bowls of wrath occur after the seventh trumpet is

blown. I'm not going to go into the seventh trumpet, because it deals with the two witnesses which proclaim the coming of Jesus' second coming. I'm also not going to go into the rapture of believers; since there is much debate and it really doesn't matter when doomsday survival is concerned. I say this because if the rapture occurs; and you have accepted Jesus in your heart and are alive at the time; and are taken up to heaven; then you don't need a safe haven. I highly recommend you thank God for <u>your eternal mansion made of the richest stones, metals, and gems.</u> You will never be foreclosed on or told to pay a mortgage and taxes.

But if you believe there is a rapture, it will happen after the two witnesses are killed by the Beast, and Jesus comes down to destroy the Beast and his armies. Hence the bowls:

The Bowls of Wrath

(Revelation 16: 2-4) **²So the first angel went and poured out his bowl on the earth; and it became a loathsome and malignant sore on the people who had the mark of the beast** and who worshiped his image.

³The second angel poured out his bowl into the **sea, and it became blood like** that of a dead man; and **every living thing in the sea died.**

⁴Then the third angel poured out his bowl into the **rivers and the springs of waters; and they became blood.**

¹⁷Then the **seventh angel poured out his bowl upon the air,** and a loud voice came out of the temple from the throne, saying, "It is done." ¹⁸And there **were flashes of lightning and sounds and peals of thunder; and there was a great earthquake, such as there had not been since man came to be upon the earth,** so great an earthquake was it, and so mighty. ¹⁹The great city was split into three parts, and the

cities of the nations fell. Babylon the great was remembered before God, to give her the cup of the wine of His fierce wrath.

[20]And **every island fled away, and the mountains were not found**. [21]And **huge hailstones, about one hundred pounds each, came down** from heaven **upon men**; and men blasphemed God because of the plague of the hail, because its plague was extremely severe.

The end result of the Earth is that it will not be around because it will be completely destroyed by just about everything from fire and objects from space, to storms, earthquakes, volcanoes, and plagues; death itself of every living thing. Other books in the bible describe particular aspects of how people will be dead and animals feeding off of them, how people will melt in the end times, or how people will go against their own blood relatives. No matter how you look at it, the end result is total destruction because of all the events that will occur, not just one all encompassing super volcano, meteor, or whatever other single event you care to imagine.

(Revelation 21: 1) Then I saw a **new heaven** and a **new earth; for the first heaven and the first earth passed away, and there is no longer any sea.**

The saying that there is no sea anymore refers to the sea which was there when God created everything in the beginning. Sea meaning space or existence itself if you want to look at it as many people do when they think of Star Trek or navigating through outer space with a spaceship, as a ship at sea. My point is not that you have to believe that space itself will not exist the way we know it, but that the Earth we know is doomed according to the Holy Bible/God, and He will make a new Earth for us to live on to include new eternal bodies. How does all this help me survive. Well it helps you see hope if you believe in God, and if you don't well it gives you something to ponder, or ignore and face science or nature as most unbelievers would term it. In which case the Earth has many years before it is destroyed with the expansion of the Sun into a Red Giant star. So you should continue to survive as I have prescribed looking out for the new generations of grand children and not just your personal survival.

I want to say this so you understand that I am a professing Christian; but am writing this book for many reasons. Any Christian believing in eternal life with Jesus and understands that once saved is always saved, should be secure that death is not the end of life. It is a physical death on this Earth, but life goes on because you are a spirit. Whether it is in an eternal hell or heaven is what is up in the air for those people who don't know the good news of Jesus' salvation for us or won't accept it. Looking at it from a logical point of view, I personally would create or join a prepper group, not because I want to live forever or to see more generations for the sake of survival, I know I will live forever. I would want to survive in this Earth so I could spread the good news of Jesus saving us for as long as possible until Jesus comes back to Earth to separate the good and evil, and set His eternal throne on the new Earth and new Heaven.

I didn't give my life to Jesus when I was a small child; in fact it wasn't until 1993 when I was saved by Jesus. Well into the years of living a life full of sin as an agnostic, sort of speaking. I could have died several times since the age of three when I for some reason told my brother to point the pistol he found at a wall instead of me and my head. To the time I flew on a plane that crashed on the flight after mine when I went to Egypt with the military. To several times on the road, jumping out of airplanes, and also playing with fire and many diseases I could have contracted in my time of sinful philanthropy. I'm sure there were many people who have come close to death many times in their lives, but caulk it up to dumb luck or personal superiority, or maybe even attribute it to God, thinking God is only suppose to provide a miracle when things go bad.

I believe that you still have to do something in this life in order to do something. I believe in grace and salvation by faith; not works. What does this mean; there are some things God requires that you do. One is you believe that Jesus as God and Man took your punishment for <u>all</u> your sin past, present and future; which opens the flood gate of grace where all you have to do is rest in Jesus. Speak and it will be done. Look at Jesus and you will not be controlled by sin or the world which is a falling world. If the world were not falling, it would not be decaying, dying, and cursed. Curse meaning does not have the favor of God, or simply said, Not Blessed.

What does this all mean? We live now, not in the past or in the future. Your body and soul feels pain and happiness just like everyone else, now and it will tomorrow. We need and have a desire to survive, so why should we do it by reacting to what science and people throw at us. If you believe in God, I would also add Satan; prep if that is what you want, love your neighbor, not because you are told or expected to, but because you can because Jesus has freed you from captivity. Survival is a mindset; a group of beliefs, so take what makes you or who you think you are and prep for the worse.

Like I said earlier, this book is for those who want to prep and survive a doomsday scenario, and if God is telling you in your spirit to prep, then I would listen to Him and prep. If you are of the non God believing or of little faith inclination, then like I said earlier, if you logically or emotionally feel you should prep and try to survive a doomsday scenario, then go for it with zeal and stay motivated even when you feel you are going backwards.

Chapter Fifteen

* | * | *

SATISFACTION CAN SPELL DEFEAT

No matter how much you prep, there is always going to be a point where you will have achieved a victory or many victories. I didn't say failures, because you will have those too, but the victories are ones that will cause you to think you are close to completing your task or surviving the crisis which hasn't happened yet.

From a worldly/objective viewpoint, if the biblical accounts are correct, then there is no amount of prepping or resources that will keep you living during the pouring of God's bowls of wrath after the seventh trumpet because the Earth will be totally and absolutely destroyed. Other than that, just hope you prepped right and hope you are at the right place at the right times.

Aside from this, the survival of things, excluding total destruction, is very high if all or most of the things mentioned in this book are performed. Prepping is a lifestyle for many people, or a hobby for some; but one thing which can cause total failure is the idea that your prepping measures are finally completed or there is nothing else that can be done to improve your survival chances. I know that when you read that sentence, you might be thinking to yourself that's not me. If it's not, then you are doing great. Otherwise, you should never be saying to yourself there is nothing left to do so let me relax. Do not confuse this with "I need to rest," because resting is something that you need to do constantly.

A ten-fifteen minute rest break every hour is nothing to be ignored or placed last on the priority list. Your break time will be the time you get nourishment of liquids if you haven't been drinking regularly; and your time to look at the work performed and step back from it. If you have been doing something wrong, the chances are you will spot it in your break time. Make it a habit to analyze things when you are resting or not distracted. If you think that taking breaks every hour is too much, you have not done your research which points to quality breaks produces quality work and more productivity. The problem is most people attribute rest breaks to laziness of someone (or more than one person) who takes too long of a break or more than one break in an hour. Depending on your workload, a five or ten minute break every twenty minutes might be best. Many people also think that talking smack or gossiping is associated with laziness or goofing off in a break. This can lead to it since people don't track the time elapsed when goofing off. Your group will do great if all your members understand and utilize rest times properly. Laughter and good conversations make breaks productive.

Each member must also follow orders that the leaders dictate or suggest. I'm not saying follow blindly, if there is a situation where a command is unsafe or clearly going to injure a person or animal for no justified reason, then you should challenge the order. Otherwise, the leader or subject matter expert should be followed. People that think they have a better idea and stand on it even though the leadership has considered it and then decided against it are not helping matters. There were many scenes in Doomsday Preppers and Doomsday Castle where individuals went off by themselves doing what they thought was better or faster. The thing is not only are they slowing down progress, the defiance to authority/leadership strangles future progress and teamwork. There is no "I" in teamwork, and when you have an "I" attitude, the individual usually thinks he/she is always right; which will cause many discipline and knowledge problems. If you follow orders out of a desire for the group and getting things accomplished you should never feel like there is nothing left to do <u>or learn</u>.

All of this information about breaks and following orders to help the group is a mindset which needs to point towards a feeling like you have not completed your tasks, goals, or plan for the day if the leader or trainer says you haven't. The leader is the one who needs to set the rate of progress and if done right, the group won't get too stressed in trying to meet an unrealistic goal for the day; or the group won't waste time doing something that is not productive. An example of something not productive would be to have a group meeting every single time you think the group needs to hear you complain about things instead of going around and spot checking/correcting issues.

I talked about some things to do with limited resources; but what if you have extensive resources? It would be great if you had almost unlimited resources. You could create a self-contained underground facility on the side of a mountain, with all the equipment to protect yourself from within the facility against hostile heavy armor or aircraft. You would have a twelve adult group who learned all the expert areas to include piloting planes and helicopters. You would make cache sites in specific locations where you can create an air strip and take out your cached cargo plane, and various helicopters, like a Coast Guard Seahawk. You might be able to get military weapons and munitions, but it really doesn't matter, because if you have enough money you can create things which come close to it. You can create a helipad and vehicle room that can contain a helicopter, dozer, recon vehicle, and a few other crafts so you can get down from and return to your mountain safe haven. You will coordinate with and find preppers that should have a good chance of surviving a worst case scenario. In the end, you must always be learning new things and never think there is nothing left to do.

I didn't talk about many things that could be considered common sense or tidbits of information which will make things easier in your prepping, but like I said from the beginning, this book is not your single source library which you should compile on all subjects. Things like putting electrical outlets at adult eye level instead of near the floor in an underground room so if the room is flooded you have time to react without a high chance of being electrocuted, is probably not going to be written down in most survival books, but it is up to you to see the benefits in such information and recommendations. Many things are complicated and many of those things require hands on or practical common sense training. An example is people are aware of a teamwork issues when they are moving and shooting their weapons. In Doomsday Castle, there was a good visual example of this when the brothers and sisters were taught about suppressive fire and movement. You can tell that when there was individualism, it didn't work at all and both people storming the castle would have died and did die by paintballs. When the other pair of siblings tried it, they did a lot better because they worked together. It was all good for the training objective to get people to work together as a team and learn some degree of tactics and technique. But here is where reality comes in. Even in the military there is this problem. The problem is the people in the castle were training with paintball rifles. The military trains with blank rounds and a few train with plastic bullets. But you see where I'm going with this. The military trains with live rounds, but only fire at fake targets. Obviously they can't really train by firing live rounds at real people. So my point is this: The expert instructor should have from the start taught the siblings that taking cover behind a bush or five inch thick tree is not going to stop a real bullet from hitting you, In addition, paintballs apparently don't do enough damage/pain to make you duck your head behind a rock or very thick tree to keep from getting hit. This is one of several reasons plastic bullets were developed.

In the end, one pair of siblings took away that they needed more work, while the other pair felt they had accomplished the task so much that they never thought about needing major improvement; just minor improvement. So what happens is - if they train like this, then they will react as they were trained. I'm sure live rounds coming in your direction will break your habit of standing in the open or behind a bush; but sometimes it will be too late to learn a fatal lesson. Always analyze your training, and hopefully your subject matter experts will train you correctly from the beginning and not let you start bad habits or learn a weak foundation. If you have a mindset that there is always something that you can improve, you will increase any survival prepping challenge.

This goes along with fratricide. The siblings were being told not to walk in front of someone's line of fire. Military personnel learn the hard way during MOUNT operations, which is training how to fight in a buildup area like buildings, rooms, towns, cities, sewers, etc. The trainees would get hit on the back of their Kevlar helmet with a wooden stick by the instructor so they felt it and understood that they had just walked into someone's line of fire. The hit they felt on the back of their head could have been a live round and it makes you really think. It was obvious that the siblings in Doomsday Castle didn't have helmets, but maybe the father could have shot a person in the back with a paintball or a plastic low powered bee-bee gun. Either way, my point is never settle with just words. I told you that hands on experience is usually the best and fastest teacher. You can tell someone all they need to know how to put someone in an arm lock with your legs, and even show them; but the person will not really remember or know how to do it until they actually perform the hand to hand maneuver.

There are many reasons you want ample space in your safe haven, and one is so you have options when things hit the fan. Never settle for a small space if you know you can get a larger space by moving things, gaining more resources, or if time permits create space with existing resources.

Last note: There are many preppers out there, and hopefully there will be many survivors after the doomsday crisis abates. It is up to the leaders and group members to increase their chances of survival. Military resources will be out there and even in a super volcano, there is a good chance military forces will survive. If you have access to military equipment, get it; don't settle for all civilian or homemade equipment if you have access to not necessarily high tech equipment, but military equipment which is usually very durable and effective. You don't have to go crazy and get everything Kevlar for example, but I would recommend you get Kevlar helmets for use in training and in the actual execution phase. You don't have to get automatic grenade launchers, but an M-203 or similar grenade launcher will do the job. I didn't go into depth at all with filtration systems to include gas masks. The reason is there is so much information out there and I recommend you research it and decide what you need. I would say you should start with gas masks that use filtration systems (filters), and masks that operate from tanks, like the ones firefighters use. Other than that, your atmosphere and water systems should not rely on outside air or water.

There is always something you can do when it comes to prepping against a doomsday scenario; wisely pick the direction you want to go, and move out and conquer with perseverance.

Chapter Sixteen

* | * | *

SECURITY IS NOT A BLANKET

S ecurity is not the people you call to help you in the mall to call the police. The security I'm talking about is not a security blanket to make you feel safe or courageous. I placed this chapter second to last because it is something that I hope will stay in your mind after you read this book. I mentioned that it takes a squad of twelve professional Soldiers their entire attention to pull 360 degree 24/7 security at any open or rugged location, simply because in the worst case scenario the squad would place two people facing opposite directions keeping an eye out while the other ten sleep and perform other duties in the middle. There will be a shift schedule and sorry to say that the squad will not be able to perform effective security for an extended period of time (week or more). A platoon (average 36 personnel) could pull security for a longer period, but even they would have a hard time after a while. In addition, early warning systems are placed out in the perimeter which is as far out as weapons can reach. Outposts are also placed out in the distance (usually 150 meters out), but I highly recommend you never do that because you don't have the manpower and can do it simpler with remote cameras.

If you think that placing traps out there in the distance and hope they warn you of an approaching hostile force, and you don't have eyes on the traps or obstacles; then you don't have any security or warning systems. Seeing and knowing what is around you and inside your safe haven is what makes security effective. Security is not just being able to shoot at an incoming horde of marauders or rioters. That is called an offensive/defensive operation in your security plan. Security is detecting all alien movement or activity coming into your perimeter and safe haven. Alien; meaning anything that is not your group members or your animals. It could be other preppers, neighbors, unknowns, stray animals, hostiles, (dismounted or mounted), or objects like a tidal wave, fire, smoke, etc. Once you have detected whatever it is, your security measures should be able to identify it enough to be able to decisively take lethal or non-lethal action. If for some reason you have a situation where people you don't know but allow them to enter inside your perimeter due to wanting to help them or vice versa; then your security plan would look for hidden weapons, agendas, and eliminate all leaks of intelligence. Your internal security will help you not in order to catch someone red handed in your safe haven (hopefully your handpicked group is true to their integrity), but it's in case an enemy gets inside your safe haven, and then they can be tracked and eliminated before they do too much damage to your group.

I will start with perimeter security measures and work my way inside the safe haven. Depending on your doomsday scenario, your security must cater to it. If you have a large farm or crops, animals (livestock), or both, then you will have a large piece of ground to cover. Your perimeter must extend beyond your owned property. You might not be able to place cameras or traps/obstacles on land that's not yours, but you can look out into land that is not your property. Cameras are the best tools to do this. However, remember all things need backups, so you should have two types of cameras, linked to motion sensors, non electrical systems like periscopes, binoculars, or even a telescope. Of course binoculars and telescopes require a person behind them. I mentioned dogs as early warning. I highly recommend you use the dogs as warning, but don't expose them to people unless you have too. In this case don't let people know you have a dog or dogs which is hard to do if they bark like crazy and you have no sound proofing. If you are underground, you could place microphones outside with a sensitivity to that of a dog or better. That way the dog can hear it inside and warn you. Just a thought, but dogs can also be used against you which I will explain in a little while. My point is that early warning systems can be used to deter or warn you to look in an area so you can get visual confirmation on the target(s).

If your scenario is such that you are in a closed in neighborhood; then I suggest you get with neighbors to link cameras and monitor all outside ground around every structure in your block. If your safe haven is in the middle of nowhere, I highly recommend you clear out up to 1,500 meters where your cameras can easily cover, day and night. If your site is out in the open or plain to see, I would strongly recommend you create a solid wall the height of four feet and no less than two feet thick along the entire perimeter. It might be used by the enemy for defense, but there are measures you can take to toss fire on the other side of the wall, and can also place hidden motion sensors on the inside of the wall which will help you in alerting you where to focus your cameras or weapons. The notion of a clearing 1,500 meters out is to catch any enemy coming to you out in the open. Your .50 Cal weapons for defense can reach further so your security should cover beyond the wall as far as the weapons can reach. In addition, your land beyond the 1,500 meters should or ideally go downhill for at least 2,000 meters. You can place cameras up high from your location and see down the hill, you might not have full view of everything, but enough so you should be able to fire indirect weapons in this downhill slope. If you are the enemy you will not be able to get any direct weapon systems on the safe haven unless you come up to the wall, which should be booby trapped. The rule of thumb is fighting uphill is always going to hurt you. Fighting downhill is always to your advantage.

Smoke could be used to screen the enemy movement as they move on the edge of your perimeter or on the attack towards your safe haven, but that is where cameras that operate on thermo signatures will eliminate smoke or other screening methods. If thermo systems are not part of your resource pool, you can have fixed firing points or settings on your mounted weapons, so you can fire at any particular location around you, and know you are hitting that area even though you can't see a thing. If you watch the movie "Kingdom of Heaven", the hero placed painted rocks out on the battlefield which told him distances in the middle of flat rocky ground, Using the painted rocks facing him, he could direct the catapult teams so they adjusted range settings and hit the targets more effectively as they came in closer to the city. The military uses range cards which tell the firing point all the information they need in their sector of fire. The information the cards have are ranges to fixed objects like trees, dead space, which is an area direct weapons cannot hit like a trench, or the back side of a wall or building, and it gives the firer a good picture of his/her fields of fire. Note: Your cameras will do most of the work when it comes to seeing the battleground, and the people behind the weapons will ensure enemy targets are engaged.

I mentioned that dogs can be used against you, and so can other things like traps or obstacles that are not being monitored. If I were a roaming prepper with the intent of plundering anything that crosses my path I would do the following: If there are traps or obstacles that are not being monitored I would steal the trap or move it. If anything, I would disable all the traps that are not monitored. When or how would I know if they are not monitored? Experienced trackers and Soldiers will not have too much trouble in finding your traps. If they are early warning traps then all they need to do is back off. Move away and see if you respond by going out to investigate. If you have a farm with animals, they are simple targets to find, and traps will be expected near the wire. I would order my group to make a hole in the wire so your animals get out and come to me, instead of me invading your land to get the animals. If it is a crop I would order my group to destroy the wire or fence around your perimeter. In essence the goal is to destroy your perimeter. Note: it is hard to destroy a four foot tall and two or more feet thick wall. If I do, it will probably attract too much attention. Oh, I will order my group to quietly destroy the perimeter. It is hard to destroy a perimeter of cameras since most of them will be close to your safe haven deep inside your property. So I would order my group to get your attention. Hopefully you will have dogs, and I will move in and get the dogs to act up every two hours. I could do this by using a dog whistle or throwing a rock – with a sling to get distance, if your dog is not trained to tell you that he/she is hearing a dog whistle, all you will know is there is something out there that the dog is warning you about. If you go out to investigate, I will analyze you/study you. I will hopefully see your security team and know numbers, genders, weapons, and get a good idea of your routines. Getting the dogs to act up every two hours 24/7 for several days will cause you to react to us because you will probably get us on camera eventually or you will ignore the dogs, and that is when we will move in closer and set out our own traps and smoke canisters. When the time is right, we will attack and your systems in place will not be as effective because you are ignoring the dogs, relying on traps that are disabled or relying on cameras that see mostly smoke. If we do it right we will wait for a respond team to come out and snipe them before we raid the safe haven. So you can see that having a safe haven on the side of a mountain over looking everything is by far one of the best defensive and easy security focused setup.

If you have a twelve adult group, you can put one couple everyday to monitor security for 24 hours. That means that each person will monitor or help monitor the cameras and security things every sixth day. A security schedule like this is sustainable for years. One person monitors while the other person acts as relief. The couple can alternate as they please as long as one person is constantly monitoring cameras, motion sensors, and other surveillance equipment. Note that the person can do homework, eat, or other activates that don't distract the monitoring too much. If you have a very large piece of land to protect and it is covered by many trees, dead space, and other things people can hide behind; I hope you have a large group of people. This is the only time it helps to have more than 12 adults, because if the situation is as I just said, it is a good idea to have a quick reaction patrol and/or roving patrol that checks the perimeter daily for breaches, signs of enemy activity, and cameras that are placed out close to the outer most perimeter. This is another reason I recommend you clear out as much as possible to as far as your weapon systems can reach.

Having said that; if you have a safe haven that is not totally underground, you want obstacles and more cameras within twenty-five meters to your safe haven structures. Obstacles in the form of pits next to walls, razor wire on the sides of walls and roofs, pillars that can stop a vehicle from crashing into a building and breaking through a structure. In worst case, keeping a vehicle far enough away so that if it blows up, your safe haven is not taken out by a car bomb. You can have ports where you can fire directly at objects next to a wall or door of your safe haven structures. If need be, you can have a kill zone just inside the entrance of your safe haven. This is your intruder compartment or entrance. At this close range is also where you can implement pepper, skunk, or CS gas sprayers. Lethal means would be fire, acid, or grenade sprayers/bomblets. Fire sprayers could be linked with fuel sprayers; a sort of flame thrower system, but not really a flamethrower, more like delayed napalm. Well you can see what I mean by close range security and options for defense. I say defense, because at 50 meters or less it is not security for detection and deterrence (by lethal force, deception, or threats); it is security for active hostile defensive measures.

The last means of security is inside your safe haven, where you will be most likely to use your hand-to-hand skills if you don't roam around outside. But before you do use your Wing Chun or Jiujutsu weapon disarming, bone breaking superpowers, I would use a pistol from a few feet to ten meter distances to take out the enemy or cripple them by shooting a leg or arm. I don't suggest you shoot to injure, unless there is a strong reason why a person would need to be kept alive if they are deadly hostile to you. If there is a hostage situation, I hope your security measures included certifying each member to hit a Nat from ten meters with a pistol or close quarter rifle. Headshots are the preferred goals when hostage takers are concerned. If a headshot is not possible, a leg shot will usually do. Chances are that a reflex from getting hit in the leg will cause the hostage taker to shoot and kill the hostage **or** he/she will release the hostage enough for a headshot or center mass body shot.

I mentioned earlier that pistols are best in close quarters; if the pistol is held correctly, someone cannot get to your pistol before you shoot them if they don't get within arm's reach of you. There are methods of keeping a pistol away from another person even if you are grappling with that person. Pistols can usually be used by anyone ten years and older, provided with proper training. So in the safe haven you have a potential of having 20 or more pistols able to kill hostile targets with a twelve adult group dynamic. There should be three or more exits out of the safe haven. There might only be two active exits and two for emergencies. The idea is that there will be or should be one or more bottlenecks/kill zones. At the entrance or in a hallway. Pistols are great for shooting around corners; hence hallways and doorway from another room. You can also have in your security plan to place fortified furniture (fortified with armor) so they can be easily moved and used as cover prior to someone storming inside the safe haven. Standard practice is to throw a grenade type weapon inside a room before you enter it. So your safe haven should have grooves on the ceiling so if someone throws a grenade it will bounce back. Usually a grenade is thrown up at the ceiling so it bounces around the room and is not blocked by normal furniture on the floor. Well as you can tell just by what I am talking about, it can become very ugly and bloody if a hostile gets inside your safe haven and starts shooting. Your early warning and distance security measures is what keeps all these bad things from happening.

Back to the preppers that were in the middle of sparse desert. Security by two adults and seven young children should not be an example to follow. Cameras are always useful and should always be used for security as a top priority. You might think that what happens if there is no electrical power. I mentioned earlier that in order to survive most doomsday scenarios, electrical power is a must, not a desire. An EMP will not permanently take out your energy source if at all, if you were following my recommendations for insulation, backups, and independent power sources. You can operate many things from battery power, and you should have a battery system able to keep things running for a day or two, should you lose solar power, wind, water, or generator power. It would be best if you could have internal water and geo-thermo power, but that takes more resources to make than solar, wind, or outdoor water power sources. Remember that all your cameras should be closed circuit and have battery backups.

I didn't go into weapons primarily because people will use the weapons they like and also have access to. Ideally, the weapon of choice should be military automatic .50 Cal or 30mm miniguns mounted on remote control platforms which can be controlled from inside your safe haven. Long distance multi-missile launchers, mortars, or howitzers for indirect fire weapons; minefields; flamethrowers for close range which most of these are illegal for you to use by the way, and pistols and shotguns for short range encounters. The type of rounds and amount of ammunition is always something to consider, especially if you are on the move or plan to move in the future. In the overall picture of security, being far away from people with very limited access will probably be your best overall security measure.

Strong leadership is a must in good security. The strong leader is not the one who brings down the hammer on anyone who needs motivation or discipline. The strong leader is the one who has earned the respect of the followers to the point of them willingly to blindly follow because they know the leader is ordering them towards success. One way to get the group into this area of having strong leadership is for all the adults to be strong leaders themselves. In the televisions series, The Walking

Dead, there are a lot of group dynamics of leadership and followers in a doomsday environment. In many cases the leaders and groups have to deal with people who plainly put - were idiots, psychopaths, ignorant, or criminally minded. The new people the groups meet and are not zombies, for some reason are either overwhelmed with the situation and act stupidly, or they want to commit crimes thinking they will get away with them or think that the actions they take will not lessen their chances of survival. If everyone understands the leadership decisions that they might or are performing, the chances are that the group will survive because of the overall leadership skills in the group and not just one person. The decisions taken for security will have to be leader led, and cannot rely on battle drills alone. The followers need to know how to lead and in turn, how to follow.

Many preppers who have not thought about the possibility of an event from happening without warning will not survive because like I said earlier, "location, location, location" will in part, determine if you will be in or far away from your safe haven when the event happens. A meteor or super volcano will not give much if any warning, so keep your survival scenario in mind. It is a good security practice to live in your safe haven, and not depend on having a site to go to from your normal residence which is more than ten minutes away. Survival should be focused on just that, and if it needs to become a lifestyle and permanent change of address, then that is what you should strive for. Your security will greatly increase if you are always present instead of depending on cameras and alarm systems in protecting your assets or safe haven while in the prepping phase.

Last note: Even if the doomsday even has abated and you and the other preppers are the only ones alive on Earth, security measures will save you from animals or dangers in and around your land. Security never quits because dangers will never quit, mostly in the form of other people.

Chapter Seventeen

* I * I *

LAST NOTES

I wrote this book in response to a suggestion one of my lifelong friends gave me, who is a doomsday prepper. I was born in Bogota Colombia, and moved to Maryland when I was six years old. My father worked in Washington DC, but the work situation changed and we moved to Miami, Florida while I was in the middle of sixth grade. I basically spent elementary school in Maryland and junior high school in Miami. I hated reading, and it's ironic that my favorite subject in school was math, and least favorite was English. Even though I was in honors math, science and English, I elected to finish ninth grade and drop out of school. I got my GED once I turned 16, and went straight into Miami-Dade community college. After completing 30 college credits, I turned 17 and a month later enlisted in the US Army as an Airborne Infantry Ranger. I spent six years as an infantryman and got out of the Army. I ended up in Louisville Kentucky studying at Boyce Bible School, and received a Bachelor's degree in General Studies with a minor in Psychology from Indiana University. I was planning on attending the Southern Baptist Theological Seminary to eventually get a PhD in World Religions, but instead I enlisted back into the US Army in June 1994 as an infantry Soldier, stationed in 25th Light Infantry Division. I attended Officer Candidate School two years later and became a Military Intelligence Officer specializing in Counterintelligence. I retired from the US Army with the rank of Major in April 2008. I have published five books, four of which are science fiction novels of an eight book superhero epic series.

I introduced my biographical snapshot information in this last chapter so you can see that the information and recommendations I have written in this book are not wild conclusions or opinions. My experiences, knowledge base, and a lot of research is where the conclusions, facts, and opinions come from. Like I said from the beginning, it is up to you to do research if you doubt anything from any source, and make an educated decision on what to believe or disbelieve. It is sort of like going around trying to find proof and people that support the claims of an advertised item.

I researched the, at the time (2012), wonder walking shoes that were cut off at the front and back (heel cups) supposedly giving the walker the ability to not strain the joints of the leg and feet while walking because of reduced impact when placing the heel of the foot to the ground. I found many people saying that they were good shoes, but also that they were very unstable to work in. Apparently many nurses got the walking shoes and many complained that their feet hurt, and stop hurting once they stopped using the shoes. There was a man who collected research and put it in simple terms on how the shoe was good if you walk straight but not for extended periods, but since it was cut off at an angle on both ends, the amount of surface contact with the ground was very small and made it unstable. The body and feet naturally compensated and so it caused the joints in the feet and legs to be stressed. The man wasn't a doctor, but half of his sources were from doctors, and the other sources were from people who actually wore the shoes. In light of the research, I bought myself a pair of comfortable running shoes with cushioned soles. I walk an average of four miles a day and have never had issues with painful joints or anything. If I would have followed the majority of what was advertised, I would have bought the shoes and probably wasted my money.

If all fails, you can always get things to experiment for yourself which can become costly in the long run; but at least you will know who was right, half right, or completely wrong. There is a measure of educated decision making in all of these things and that includes research conducted from mixed sources. I thought about some doomsday scenarios like zombies. But after looking at the research and books put out there, the only real conclusion I agree with is the idea that a biological event can or will happen that will contaminate people and they will not become undead per say, but they will be crazy or infected for a lack of better words. If you have seen the 2010 movie, "The Crazies" then you can understand that an epidemic of a virus or bacteria that can cause people to go psycho or infected with a deadly and contagious disease, then yes the term zombies could be used. Is it going to be like in the 2013 movie "World War Z"? Not likely, because the laws of biology would most likely not cause a person to transmit a virus or bacteria only by a bite. I suspect as I talked earlier about poisonous contamination of water from something in outer space could start an pandemic of psychotic or deleterious people going around killing people; but not true zombies as we would see as animated undead as alive. Having said that; the measures taken to the extent of an underground, fortified, and well protected (maximum lethal measures) safe haven will stop zombies too.

There was an episode in a show called "Zombie Apocalypse," where a discussion started on what you would do if your son or daughter becomes contaminated. The group argued that they would not be able to kill their own offspring, or that they could and it was clear that there was mixed emotions in the group. I said earlier that survival is a mindset, not just a physical thing to achieve. Small water filter straws are great, but the mindset should be that the straw is a last measure backup. The primary and secondary are having a self contained water source and self contained filtration system in a large scale. The rule of thumb is to not get exposed to or be near contamination situations or other lethal sources. In the case of having to struggle with your love for your children, you should be prepared to eliminate the imitate threat for the sake of the group, but the real question should be how to keep from getting into that situation to begin with.

I want to address companies like Deep Earth who construct bunkers for preppers. The bunkers they make are impressive, but not all of them are good or practical for your desired outcome. They can try to cater to your needs, but in the end, they only supply you with what your wallet can come up with. Most of the bunkers are constructed out of metal which is how you need to make it to construct it and then place it, ready built, into the ground. I talked about the layers of suggested materials for your safe haven, and whatever route you take, don't settle for a premade bunker and forget about all the other aspects of prepping and surviving. Look at all your options and if companies and other prepper groups you research are for you, go for it.

I want to point out that many tidbits of information that I could have gone into, I didn't because it would take too much time and may not apply to you. Like I said from the beginning; your prepping survival skills come from manuals, step-by-step instructions, and hands on training on top of research you have or will compile. Hands on experience is one of the best ways to learn, and I cannot give you that in one or several books. I did try to show you the mindset you need, and methods in planning, focusing your efforts, and lastly looking at a future instead of just the work in front of you.

I can't stress enough the need for backups on everything. This includes knowledge and experience skills. All adults must be proficient in being leaders. There needs to be two doctor or nurse level experts in the group. Depending on the crisis, chances are you won't be able to call time out and take your child or adult to the family doctor or hospital. The same goes with needing to have veterinary skills enough so you can care for your dog or livestock instead of thinking there will be a vet near where you live. Stocking up on medicine is fine, but you need to know how to use the pharmacy you created in one of your rooms in your safe haven.

I talked about location, location, location... and unfortunately I must inform you that resources are sometimes not going to be in your favor because of the location you choice to pick. Not every state or location will allow solar cells, rain water collectors, or wind mills for example. Certain weapons are illegal in certain states. So you must do research and find out if water collectors are legal for your location if you plan on rain as a water source, because if they are not legal, then you must come up with an alternate way of gathering water or be prepared to be fined by the county or in worst case scenario go to court because a statue is not being met. It also doesn't have to be the state or county, it could be a Homeowners Association that will try to give you problems. My strong suggestion is if there is a HOA don't get property there, or ensure the HOA is not going to give you problems in the future by becoming the head of the HOA, or something similar to it. Aside from the HOA, the county and state have more of a claim to fine you and put you in jail for breaking building codes or statues. The rain water collectors for instance are illegal in some states because they are a health hazard since not everyone takes care of them, hence there is contaminated water and insects/birds spread disease to other animals and humans. Solar cells are said to be too difficult to dispose of and are a hazard to the environment; likewise wind mills; but I personally think the later two are illegal in many places because of politics and money. Electric companies are the big pushers for restricting solar and wind power, since they will lose a lot of business, so they think. In many places, if you get extra energy from your solar or wind power system, you can sell that extra power to the electric company and actually make a profit for having your own power source. The electric company gets your energy and gives you money, but in actuality they are selling it to other customers and making a small profit for themselves on the side.

There are also other considerations you need to know like snow falling on your solar panels which if not maintained or positioned properly will be damaged from the weight of the snow. My point in all of this is you need to do research on all aspects of your location, materials, equipment, and people to make sure you won't run into brick walls which usually come in the form of legal court issues or bad relationships.

A last topic I want to address is there's the concept that an alien invasion would cause a doomsday event. If there are aliens with the capacity to travel through space, they probably will either; leave us alone, kill us all from outer space or on the ground, or boost us into higher levels of technology. If the worst case scenario is that they will kill us or enslave us, surviving in a safe haven will more than likely only delay the extinction of humanity sort of speaking. In this "Independence Day" or "Falling Skies" situation the only survival strategy is for everyone with the capacity to engage the enemy and fight is the only survival method that will work or fail. War on a planetary level against an alien race is something that you can prep for, but it won't help much if everyone else is not on the same sheet of music; which fortunately for us, most people are prepared for war because our history and current status on fighting is constantly being taught to our children and ourselves. The movies like "Battle: Los Angeles" is a good example of how people come together to fight. This is one reason I didn't address an alien invasion because if you survive by yourself or in a small group, it won't do you any good because the threat will not leave the planet unless the majority of people in the world can survive by fighting. If you for some reason think that the aliens will die from disease or some "War of the Worlds" type of situation, then you are kidding yourselves. Chances are that we will die from alien diseases or bacteria.

I hope this book helps in your endeavor to survive events which frankly speaking are things that with God's favor you will not witness; or if you do, it will be to survive for His purpose to save others by spreading the good news of His salvation for us. Time will tell, and you need it to prep.

May God's favor be on you and His peace comfort your heart and mind.

All references to biblical scripture were taken from the NASB. (1990). The new American standard bible; open bible, study edition. Thomas Nelson, Inc. Nashville, TN.

www.ingramcontent.com/pod-product-compliance
Lightning Source LLC
Chambersburg PA
CBHW060013210326
41520CB00009B/875